PREPRINTS

Charleston Briefings: Trending Topics for Information Professionals is a thought-provoking series of brief books concerning innovation in the sphere of libraries, publishing, and technology in scholarly communication. The briefings, growing out of the vital conversations characteristic of the Charleston Conference and *Against the Grain*, will offer valuable insights into the trends shaping our professional lives and the institutions in which we work.

The Charleston Briefings are written by authorities who provide an effective, readable overview of their topics—not an academic monograph. The intended audience is busy nonspecialist readers who want to be informed concerning important issues in our industry in an accessible and timely manner.

Matthew Ismail, Editor-in-Chief

PREPRINTS
THEIR EVOLVING ROLE IN SCIENCE COMMUNICATION

IRATXE PUEBLA, JESSICA POLKA
AND OYA Y. RIEGER

Copyright © 2021 by ATG LLC (Media)
Some rights reserved

This work is licensed under the Creative Commons Attribution-NonCommercial-NoDerivatives 4.0 International License. To view a copy of this license, visit http://creativecommons.org/licenses/by-nc-nd/4.0/ or send a letter to Creative Commons, PO Box 1866, Mountain View, California, 94042, USA.

Published in the United States of America by
ATG LLC (Media)

DOI: https://doi.org/10.3998/mpub.12412508

ISBN 978-1-941269-47-3 (paper)
ISBN 978-1-941269-48-0 (ebook)
ISBN 978-1-941269-50-3 (open access)

http://against-the-grain.com

CONTENTS

Acknowledgements vii
1. HISTORY OF PREPRINTS AND CURRENT LANDSCAPE 1
2. PERSPECTIVES ON PREPRINTS: ADVANTAGES AND CONCERNS 11
3. SOCIAL MEDIA, COMMENTARY AND OPEN REVIEW OF PREPRINTS 20
4. TECHNICAL INFRASTRUCTURE 29
5. STAKEHOLDER PERSPECTIVES 34
6. PREPRINTS AND PUBLISHING 41
7. PREPRINTS AND RESEARCHER ASSESSMENT 50
8. PREPRINTS AS AN OPEN SCIENCE TOOL 55
9. CONCLUSION AND OPEN QUESTIONS 60
References 65
About the Authors 75

ACKNOWLEDGEMENTS

We would like to thank Matthew Ismail for the opportunity to contribute this Briefing and for his helpful and supportive comments on different versions of the manuscript. We thank Mario Malički for his thoughtful review of the manuscript. We also wish to acknowledge our colleagues and the members of the ASAPbio Community with whom we regularly discuss preprints and their role in the ever-evolving science communication ecosystem; special thanks to the researchers in the ASAPbio Community for being a continuous source of inspiration.

CHAPTER 1
HISTORY OF PREPRINTS AND CURRENT LANDSCAPE

'We look at the present through a rear-view mirror. We march backwards into the future.'

—Marshall McLuhan

WHAT IS A PREPRINT?

A preprint is a scholarly manuscript posted by the author(s) to a repository or platform to facilitate open and broad sharing of early work without any limitations to access. The preprint content is generally similar to a manuscript submitted to a scholarly journal and may be posted before or in parallel to submission to a journal or independently of whether the paper will be sent to a journal. Typically, after a basic screening process, the manuscript is posted on the preprint server within a few days of submission, without peer review, and it is made freely available online. Preprint servers do not require copyright transfer, allowing the authors to retain copyright and to post the paper under different licenses that enable others to reuse the work (permitted uses will vary depending on the license).

The main appeal of preprints is that they allow authors to share their work openly, early and rapidly, with a much shorter turnaround than is required for publication in a peer-reviewed journal. With the preprint model, authors can control the dissemination of their work and share their research with the scientific community as and when they are ready to do so without being limited by the processing timeline associated with formal publishing (Chiarelli et al.,

2019a). In addition, researchers may use preprints to share work with the community that they may not intend to publish in a journal. Publication trends for preprints show that in some disciplines a majority of preprints are eventually peer reviewed and published in journals (69 per cent of physics preprints (Xie et al., 2021) and 66.5 per cent of working papers in economics have an associated journal article (Baumann & Wohlrabe, 2020), 67 per cent of preprints in bioRxiv appear in a journal within two years (Abdill & Blekhman, 2019)).

What constitutes a preprint and the content types supported by preprint servers continues to be a topic of discussion in the scholarly community (Rieger, 2020). Repositories hosting preprints may also include postprints,[1] conference papers, working papers, reports, white papers, literature reviews, book chapters, slide decks and posters. Some preprint servers such as arXiv and preprints.org support the submission of supplementary files (at modest file sizes) associated with the paper (e.g. images, spreadsheets or program code). Preprint servers generally hold preprints in perpetuity, offering in some cases to link to a subsequent version of record when possible.

The goal of this Briefing is to discuss the history and role of preprints in the sciences within the evolving open science landscape and to share recent developments related to their uptake, review, technical infrastructure and business models. Although the discussion applies to many other disciplines, our coverage is tilted towards life sciences (particularly bioRxiv) as the landscape in this domain has changed dramatically over the last five years. As of the writing of this Briefing, there are more than 60 preprint servers representing different subject and geographical domains. As each one of them is evolving at a different pace based on adoption patterns and disciplinary ethos, rather than a meta-analysis, we will focus on specific preprint servers such as bioRxiv, arXiv, Research Square and SSRN.

EARLY EXPERIMENTS AND FIRST PREPRINT REPOSITORIES

While the use of preprints is a relatively new phenomenon in the life sciences, experiments to allow researchers to disseminate their work early and via channels outside of journals go back to the 1960s.

---

1 A postprint is also known as the 'author-accepted manuscript', the draft of the paper after peer review and acceptance at a journal but prior to the typesetting and formatting processes.

In 1961, the biochemists David Green (University of Wisconsin) and Philip Handler (Duke University) and the National Institute of Health (NIH) administrator Errett Albritton developed the idea of the Information Exchange Groups (IEGs) as a mechanism for researchers to share early work with their peers. Researchers could send their paper to the NIH which would then circulate the document through the mail to the various members of the IEG network (Cobb, 2017). Interest in the IEGs grew and different disciplinary groups were created; by 1965, over 3,600 researchers were involved and 2,500 memos had been circulated (Cobb, 2017).

As IEG membership grew, journals began to express concerns about their model. The editors of *Nature* and *Science*, for instance, both criticized IEGs, noting that the groups were expensive to run and suggesting that they were a channel to circulate work of uneven quality that circumvented the structured vetting of manuscripts. In a meeting in 1966, the editors of 13 leading biochemical journals decided that they would no longer publish papers that had been circulated as IEG memos. In light of the restrictions that this decision imposed for researchers in disciplines overlapping with some of the largest IEGs, as well as concerns over growing service costs, the NIH closed the IEG groups in 1967 (Cobb, 2017).

In parallel to the IEG experiment, communities in the physical sciences had conversations about how to facilitate the circulation of early papers. One proposal suggested the formation of a Physics Information Exchange (PIE), which would be similar to IEGs but focused on high-energy physics. The difference would be that papers in the PIE would be circulated to participating libraries instead of directly to researchers. PIE distributed weekly lists of preprints for a year as a trial and was more cost-effective than IEGs because it did not circulate full documents. Readers would instead request the preprints in which they were interested from the authors (Till, 2001). PIE was followed by the 'Preprints in Particles and Fields' (PPF) in 1969, which ran a weekly publication of preprint lists and was operated by the Stanford Linear Accelerator Center. PPF included a section called 'anti-preprints' which listed preprints that had been published in journals, an addition that sought to address potential reservations among editors (Till, 2001).

In 1989, the physicist Joanne Cohn began circulating string theory preprints via email in an effort to facilitate the transition from paper to online formats. The email list grew over the following two years, becoming increasingly challenging to manage both for Cohn and for the capacity of mailboxes at the time (Garisto, n.d.). Another physicist from Los Alamos National Laboratory (LANL), Paul Ginsparg, offered to automate the list, leading to the birth of arXiv (Garisto, n.d.). In August 1991, Ginsparg created a mailbox repository for papers and moved this to the World Wide Web in 1993 ('arXiv'). The repository, which was originally hosted at LANL, moved to Cornell University in 2001 when Ginsparg assumed an academic position at this institution. arXiv began as a repository for physics papers but expanded to other fields such as astronomy, mathematics and computer science.

Other repositories aimed at the physical sciences also started around this time. The Mathematical Physics Preprint Archive (*Mathematical Physics Preprint Archive*, n.d.), hosted by the University of Texas–Austin, was set up in 1991 as a repository of research papers in mathematical physics and related areas; the Electronic Colloquium on Computational Complexity (ECCC) was founded in 1994 at the University of Trier (Germany) as an archive of papers in computational complexity theory. ECCC aimed to provide an intermediate between the basic screening at arXiv and peer review at journals, with its screening also including considerations of novelty and interest in the submitted work ('Electronic Colloquium on Computational Complexity', 2019).

Preprint initiatives in other fields followed with the launch of the Social Science Research Network (SSRN) in 1994 as a repository of papers in the social sciences and humanities. SSRN was founded by Michael Jensen and Wayne Marr 'to enable scholars to share and distribute their research worldwide, long before their papers work their way through the journal refereeing and publication process, and to facilitate communication among scholars at the lowest possible cost' (Elsevier, 2016). Research Papers in Economics (RePEc) followed in 1997 for the dissemination of research in economics. RePEc's roots go to a 1993 initiative called NetEc, which was a volunteer effort to improve the communication of research in economics via digital media. Its founder, Thomas Krichel, described the purpose of NetEc as, '[opening] Economics to the public by improving both current awareness and access to publications and other data' (Krichel, 1997). The vision was 'to fight the division of the world into informationally rich and poor'.

As the use of preprints for the rapid dissemination of scholarly work began to expand into different disciplines, some expected that this adoption would extend into the life sciences. However, the trend did not immediately materialize. By the early 2000s, many physicists and mathematicians were regularly circulating their work and reading peers' early research via preprints, but biologists had still not warmed up to the model. *BMJ* launched ClinMedNetPrints.org as a preprint server for the clinical sciences in 1999 but this closed nine years later due to low adoption. In 2007, the Nature Publishing Group launched the preprint server Nature Precedings as an open electronic preprint repository of works in the fields of biomedical sciences, chemistry and earth sciences. However, it also failed to gain traction and ceased activity in 2012 (Cobb, 2017).

SECOND WAVE OF PREPRINTS: INITIAL ADOPTION IN THE LIFE SCIENCES

Things started to change in 2013 with the launch of PeerJ Preprints and bioRxiv, both dedicated preprint platforms for the life sciences. The open access publisher PeerJ established PeerJ Preprints in April 2013 with the stated goal of supporting authors across the publication process, from the step of creating and hosting a preprint to submitting that work for publication in a peer-reviewed journal (Binfield & Hoyt, 2013).[2] Also in 2013, John Inglis and Richard Sever from the Cold Spring Harbor Laboratory co-founded bioRxiv with the hope that biologists were finally ready to embrace preprints to 'share their raw manuscripts on a free online archive before sending them to a peer-reviewed journal' (Kaiser, 2014).

The accelerated emergence of open access publishing in the 2000s had brought renewed conversations about approaches to the dissemination of scholarly work. While biologists' willingness to start engaging with preprints is probably associated with a combination of factors, a few potential elements are likely to be particularly relevant. First, by the 2010s, most journals had adopted digital formats and researchers were increasingly familiar with online modes of communication. Another important factor was that bioRxiv was hosted by the Cold Spring Harbor Laboratory, a non-profit research institute that also operates reputable journals. This mission-driven association allowed bioRxiv to establish partnerships with a number of journals and thus mitigate some of the resistance that had been a major obstacle for previous preprint initiatives. The advent of social media

---

[2] In 2019, PeerJ Preprints decided to stop accepting submissions as its founders felt that other science preprint venue options had started registering success (Hoyt, 2019). Meanwhile, bioRxiv flourished.

platforms such as Twitter was an additional contributing factor to the adoption of preprints, as researchers started to use social media to amplify the visibility and engagement with work they posted as preprints (Penfold & Polka, 2020).

Although bioRxiv and PeerJ Preprints attracted several hundreds of submissions in their first couple of years, support for preprints in the life sciences accelerated further in 2016. Early that year, the Accelerating Science and Publication in Biology (ASAPbio) meeting gathered a group of research communication stakeholders to discuss how preprints might facilitate the communication of biological research ('2016 Meeting', n.d.). There was broad support for the use of preprints among attendees, and as a follow-up to the meeting, ASAPbio developed into a non-profit organization to coordinate and support efforts towards the adoption of preprints in the life sciences.[3]

2016 saw the arrival of several more preprint platforms. The Center for Open Science launched Open Science Framework (OSF) Preprints as an open source preprints platform both to facilitate new models of scholarly communication across multiple disciplines and to improve the accessibility of scholarship. OSF played an important role in increasing the number of preprint initiatives as it provided a hosted technology platform with a search interface covering all hosted servers (Nosek, 2017). In the same year, the open access publishing service provider MDPI introduced preprints.org as a multidisciplinary platform to make early versions of research outputs permanently available and citable (MDPI, 2016). ChinaXiv also launched in 2016 as an open repository and preprint distribution service for Chinese researchers in the fields of natural sciences, maintained and operated by the National Science Library of the Chinese Academy of Sciences ('Preprint', 2020). The trend continued in 2017, when SSRN launched the Biology Research Network (BioRN), a platform for posting preprints and working papers in biology (*SSRN Launches a New Network Dedicated to Biology – BioRN*, 2017).

Several funding agencies also announced policies at this time that allowed or encouraged the use of preprints as evidence of research productivity in grant applications. New policies announced by the Simons Foundation (United States) and the Medical Research Council (United Kingdom) in 2016, and the NIH and the Wellcome Trust the following year, signalled funder support for preprints ('Funder Policies', n.d.). In addition, several institutions such as the

---

3 ASAPbio is a scientist-driven non-profit with a mission to promote innovation and transparency in life sciences communication. ASAPbio coordinates initiatives that support the adoption of preprints in the life sciences as well as increased transparency in peer review.

Howard Hughes Medical Institute (HHMI, United States) and CNRS (France) announced that preprints would be considered in hiring and tenure application processes ('University Policies and Statements on Hiring, Promotion, and Journal License Negotiation', n.d.).

As support grew and the preprint landscape broadened, new national and regional platforms by open science advocates continued to emerge. These services were introduced as multidisciplinary portals to foster open scholarship. For instance, in 2018 AfricArXiv was launched to improve the visibility of African science by helping researchers share their work quickly and fostering collaboration (*African Scientists Launch Their Own Preprint – Scientific American*, n.d.). The AfricArXiv platform accepts preprints, postprints, code and data in all African languages. In 2019, IndiaRxiv was established to provide a national preprints repository for India ('Preprint', 2020). There are currently at least four preprint servers with a focus on research from specific geographical regions or languages ('List of Preprint Servers', n.d.).

INCREASED ADOPTION AND PUBLISHERS ENTER THE PREPRINT STAGE

The growing visibility of preprints, and their recognition as valid research outputs by funders and national institutions, was followed by an increase in posting of life sciences preprints. bioRxiv content grew rapidly, from around 1,700 submissions in 2015 to 20,000 new preprints in 2018. The number of preprints in the life sciences posted in other servers also increased, albeit at a less dramatic pace. arXiv content in biological and medical sciences went from 5,425 in 2015 to just over 6,700 in 2018; for SSRN, content in those disciplines tripled (931 in 2015 vs. 3,128 in 2018) (*Dimensions*, n.d.).

While the volume of preprints in the life sciences still constituted only a small proportion compared to peer-reviewed journal publications (1–2 per cent of the biomedical literature in PubMed in 2018 (Polka & Penfold, 2020)), the increasing support among scientists and the rapid growth in preprint deposition aroused the interest of publishers. Over the last few years, almost all the big publishing houses have explored options to incorporate preprints into their workflows, by adopting partnerships with preprint platforms, by developing their own preprint servers or associated services or by acquiring existing platforms (Schonfeld & Rieger, 2020). As of 2021, there are 12 preprint servers relevant to the life sciences owned by a publishing organization ('List of Preprint

Servers', n.d.) that offer authors the option of having their journal submission posted as a preprint. Publisher-driven services have become an important channel for preprint deposition; the server Research Square, which runs in partnership with Springer Nature journals, reached 100,000 posted preprints in less than three years of operation.

There are now over 50 preprint platforms with a disciplinary scope covering biology and/or medicine operating under different models of ownership and governance (Malički et al., 2020).[4] As integration with journals expands, we can expect the increasing trend in preprint deposition to continue in the coming years.

While adoption is increasing, the use of preprints varies widely based on disciplinary cultures. Research communities with a culture of open science practices (e.g. around data sharing) are among those that show the strongest adoption. There seems to be a correlation between uptake of open access and preprints (Severin et al., 2020). For instance, journals in the fields of physics, mathematics, astronomy and information science were the early pioneers of open access, and the scientists in those fields make heavy use of arXiv. On the other hand, uptake of preprints in the fields of engineering and chemistry has been low, corresponding to the relatively lower prevalence of open access in those disciplines. In 2018, neuroscience, bioinformatics, evolutionary biology and genomics were the research disciplines with the largest number of manuscripts in bioRxiv (Abdill & Blekhman, 2019) with neuroscience constituting almost 20 per cent of the content posted on the server (*BioRxiv Reporting*, n.d.). Preprints data in the Dimensions database show that neuroscience preprints constituted almost 5 per cent of publications in that field in 2018, while for evolutionary biology preprints reached 6 per cent in comparison to article publications that year. These percentages are higher than the average across biomedical fields which stood at 2.5 per cent (*Dimensions*, n.d.).

COVID-19'S IMPACT AND INFLUENCE ON SHARING EARLY RESEARCH

The COVID-19 pandemic had a major influence on preprinting, particularly among those communities working on pandemic-relevant research. As the severity and global impact of the coronavirus outbreak extended at the

---

4 A directory of preprint servers for the biomedical sciences is available at the ASAPbio website ('List of Preprint Servers', n.d.).

beginning of 2020, the urgency in addressing the social and public health crisis led many of these researchers to post their work as preprints for rapid and broad dissemination.

Preprints had been proposed as a mechanism to disseminate scientific findings in response to infectious disease outbreaks (Johansson et al., 2018) and their use was encouraged in the context of the 2016 Zika epidemic; while the number of preprints related to outbreaks increased, their adoption remained relatively low. Things would be very different for COVID-19: preprints began to play a dominant role early in the pandemic. At the end of March 2020, about half of the COVID-19 papers listed on the iSearch COVID-19 portfolio (*COVID-19 Portfolio | Home*, n.d.) were preprints. Although findings reached the peer-reviewed literature and the proportion declined in the following months, as of June 2021 the platform listed over 33,000 COVID-19-related preprints. Almost 40 per cent of those papers had been deposited at medRxiv, a preprint server for the health sciences launched in June 2019 by the Cold Spring Harbor Laboratory in collaboration with BMJ and Yale.

The urgency to address the pandemic prompted researchers not only to use preprints for the dissemination of their work but also to share papers at a more preliminary stage as data were becoming available. An analysis of preprints in the period from January to October 2020 found that COVID-19 preprints were shorter (median 3,965 vs. 5,427 words) and contained fewer references than non-COVID-19 preprints (Fraser et al., 2021).

There was also widespread use of pandemic-related preprints by the research community and beyond; in the initial months of 2020, COVID-19 preprints received almost 30 times more downloads than non-COVID-19 preprints, and they were regularly mentioned, in both social media and the news, at rates considerably higher than non-COVID-19 preprints (Fraser et al., 2021). Research findings shared as preprints have also played a role in informing policy as national bodies and health organizations sought to develop strategies to address the pandemic. The policy paper on the coronavirus action plan posted by the UK government in March 2020 (*Coronavirus Action Plan*, n.d.) cited early epidemiological data from a group of Chinese patients that had been posted as a preprint on medRxiv. The initial results from the University of Oxford's RECOVERY trial were shared as a preprint (also on medRxiv) and prompted the World Health Organization to call for an increase in the production of dexamethasone after the trial reported that this corticosteroid reduced mortality among patients with COVID-19 receiving mechanical ventilation (Mahase, 2020).

The increase in early sharing of research brought with it concerns about the practice of disseminating unvetted research. Some worried about the increase in the noise-to-signal ratio as journalists began reporting on preprints with yet unscrutinized findings. During the COVID-19 pandemic, servers including bioRxiv and medRxiv added new policies (Kwon, 2020) and cautionary labels to preprints to emphasize that they are preliminary reports of work that have not been certified by peer review and should not be reported by the news media as established information.

CURRENT LANDSCAPE

As the pandemic evolves and researchers who redirected their efforts to COVID-19 research return to their original lines of work, we can expect the rate of preprint posting in relation to COVID-19 to stabilize and eventually decrease, a trend that we are already observing in the initial months of 2021. However, it is likely that the pandemic has permanently altered preprint adoption in some subfields within the life sciences. The unprecedented level of attention around preprints in 2020 has resulted in more research communities becoming familiar with this approach to science dissemination. In June 2021, monthly preprint uploads to Europe PMC represented 12 per cent of the research article uploads, a remarkable increase in comparison to a 0.6 per cent proportion in June 2016. Combining this growth with broader support for preprints across journals and different stakeholders, we can expect to see a continuing increase in preprint adoption across several research fields over the coming years.

The following sections discuss different aspects of the current landscape of preprints in the sciences, especially life sciences, including preprint review and feedback, technical infrastructure, stakeholder perspectives on preprints and where preprints fit in the current broader publishing space. Where available, we discuss existing reports on preprint use, views and practices, but it's important to acknowledge that a considerable part of the existing evidence originates from research on bioRxiv use. While bioRxiv has played a predominant role in the adoption of preprints in the life sciences, the preprint landscape includes a wide range of servers that have different scopes, governance and business models, and thus, further research is needed on how perceptions and practices relate or differ across different platforms and disciplines.

CHAPTER 2
PERSPECTIVES ON PREPRINTS: ADVANTAGES AND CONCERNS

BENEFITS OF PREPRINTS

Preprints present several potential benefits, both for researchers and for overall scientific progress.

Preprints give researchers the freedom to communicate their work rapidly, broadly and when they are ready to do so. Relative to publication in a journal, preprints allow authors much more control of when and how to disseminate their work. Paul Ginsparg has noted that the benefits for an author in posting their work to arXiv are not only to 'speed up the research enterprise, but also to make it fairer, by giving global research communities equal access to the latest results' (Ginsparg, 2016).

Preprints are posted within days of submission, so can be disseminated much more rapidly than journal publications, where the peer review and editorial process can take months or even years. Publication timelines vary widely per discipline, and while some publishers can boast relatively efficient processing (Petrou, 2020), traditionally a peer-reviewed manuscript in the life sciences takes several months from submission to publication. This period does not account for the possibility of the manuscript being rejected by one journal and having to undergo a new editorial process at another. By comparison, a preprint server can provide immediate posting or take just a few days (Nouri et al., 2020), and while submissions may be rejected during the preprint server's screening process, the rejection rate is most likely significantly lower than that at most journals.

Preprints offer a number of additional benefits for researchers (Berg et al. 2016) beyond speed of publication.

Access

Preprints are made freely available to everyone. Preprint servers offer no financial restrictions on posting or reading content.[5]

Proof of productivity

Preprints are permanent citable records and provide evidence of research productivity which researchers can share with funding agencies and promotion and hiring committees. This can be particularly relevant for early career researchers since the time for students to publish their first first-author peer-reviewed paper has increased by over a year compared to the 1980s (Vale, 2015).

Preprints can provide an avenue to share findings that traditionally have been harder to place as a journal publication, such as null, negative or inconclusive results. In addition, a preprint can be a mechanism for researchers to disseminate work and ideas which they may not intend to submit for journal publication such as proposal documents or open letters (Malički et al., 2021).

Priority

The preprint allows the author(s) to establish priority for their findings. In several subfields of physical sciences, preprints are the main mechanism for disseminating work and establishing priority. As discussed in the next section, while some concerns remain about the possibility of scooping (where, e.g. another researcher/group may see the preprint and rush to publish similar work to claim priority over the findings), communities in the life sciences may evolve to a similar approach to that of some subfields of physics and recognize priority for research posted as a preprint (Vale & Hyman, 2016). As a sign of moves in this direction, several journals offer 'scoop protection' policies that extend to the day of posting the preprint version of the manuscript (Pulverer, 2016).

---

5 A few servers, for example, Cell Sneak Peek or Preprints with The Lancet, require registration to access and download content.

Visibility

Preprints can bring additional (early) visibility to the work (Xie et al., 2021). Several studies of published research have shown that posting a preprint is associated with higher social media attention and citations for the publication once it appears at a peer-reviewed journal (Fraser et al., 2020; Fu & Hughey, 2019; Serghiou & Ioannidis, 2018).

In addition, posting a preprint can also facilitate invitations to present at scientific conferences or even open opportunities for collaboration among groups working in related projects.

The preprint can also attract attention by journal editors. Several journals (e.g. *PLOS Genetics, Proceedings of the Royal Society B* (Barrett, 2018; Barsh et al., 2016)) have appointed designated 'preprint editors' who scout the latest research posted at preprint servers and invite submissions to their journal.

Feedback

Preprints allow authors to get feedback on their work. Some preprint servers provide a forum for public comments on the preprint, and scientists may also provide comments privately over email. This feedback can help authors revise and improve the paper prior to eventual submission to a journal, and this allows a broader range of perspectives on the work than do the views of a couple of scientists involved in a journal's peer-review process (Malički et al., 2021).

Considering the research ecosystem more broadly, preprints also bring potential benefits to the overall scientific enterprise:

Speed of scientific discovery

The dissemination of new knowledge can accelerate additional discoveries, and thus the rapid sharing of the latest scientific findings can benefit society. The COVID-19 pandemic has provided a clear example of a large-scale crisis in which the open and prompt sharing of information can make a difference from a public health and societal perspective.

Higher return on investment and cost-efficiencies

From the perspective of research investment, preprints can help create different types of research outputs. In the current journal system, it can sometimes be difficult for researchers to disseminate all of their work, either because the findings may not 'fit' the format of the journal article (e.g. negative results, short observations) or because other circumstances make the bar too high to invest in the preparation of journal submission (e.g. a graduate student or a postdoc moving to a different institution and no longer available for a lengthy revision process). Preprints provide a means of sharing those types of work and thus maximize the knowledge shared from the same project grant.

The sharing of ideas months prior to the journal publication can also avoid duplication of effort. If a preprint reports that a line of research may be unproductive, other scientists can adjust their work to prevent repeating that line of research. A survey carried out as part of a review of the IEGs reported 346 occasions when information circulated in the group had prevented needless duplication of effort (Univekiity, n.d.). From an economic perspective, the results from the survey suggested savings of approximately 10,000,000 USD/year (the equivalent to 74,500,000 in 2018 dollars).

CONCERNS AND CHALLENGES OF PREPRINTS

While preprints have several potential benefits, they also bring with them a variety of challenges. These challenges may provide some context for their slower adoption in some research disciplines. We will explore here some of the most common concerns raised concerning preprints.

Scooping risks

Researchers are commonly concerned about the possibility that having work available as a preprint will allow their ideas or results to be published by others before the preprint authors can do so. This would deprive the preprint authors of rightful recognition for the work. While it is not rare to hear this concern mentioned in conversations with researchers, there is no evidence that 'scooping' is common or that it differs from situations that may arise in the context of journal publication. Given that preprints are posted publicly as time-stamped records, Paul Ginsparg has argued that having a preprint provides protection for establishing credit for the work ('Preprint FAQ', n.d.). In a survey of bioRxiv users, only 1.25 per cent of the respondents indicated that posting a preprint negatively affected their priority claim for the work (Sever, Roeder, et al., 2019). In a separate survey of stakeholders across all research disciplines carried out in the summer of 2020, which asked about the benefits and concerns about preprints, 52 per cent of those who had not posted a preprint indicated that getting scooped by others was very or somewhat concerning. By contrast, among the respondents to that survey who had previously posted a preprint, only 32 per cent marked getting scooped as very or somewhat concerning ('Preprint Authors Optimistic about Benefits', 2020). This suggests that there may be a disconnect between the researchers' perception of the risk of scooping and whether this takes place in practice.

Reliability and credibility

Given that preprint servers conduct only light screening of manuscripts, it is understandable that some are concerned about the reliability and trustworthiness of preprints. A 2019 survey of almost 4,000 researchers across a wide range of disciplines concluded that preprints can improve and accelerate scholarly communication if researchers view them as credible enough to read and use (Soderberg et al., 2020). By adding indicators of transparency/openness of research content and process (e.g. links to data and pre-analysis plans, computational reproducibility) preprint servers may be able to help researchers better assess the credibility of posted preprints, allowing scholars to more confidently use them. However, the study concluded that preprint services often do not include the heuristic cues of a 'journal's reputation, selection, and peer-review processes that, regardless of their flaws, are often used as a guide for deciding what to read'.

The potential risks of distributing work that is not peer reviewed will differ depending on the nature of the research and the claims it makes. The risk for society in distributing work that is not peer reviewed is clearly higher in the context of public health or patient care. The publication of medical research at a journal often involves a careful peer-review process evaluating different elements of the study design, claims and limitations of the work. Given that preprint servers do not provide such a review process, research posted as a preprint should not be used as established clinical evidence. medRxiv has implemented additional screening checks to apply extra scrutiny to papers reporting findings that may present a risk to public health. This screening framework seeks to weigh the benefit of sharing the information immediately versus the potential dangers, and in some cases they have asked that the work undergo peer review first (*Pandemic Preprints – a Duty of Responsible Stewardship*, 2021).

The 'light-touch' screening process at many preprint platforms is focused on establishing that the paper reports research (in the structure and format expected for scholarly papers) and that there is no inappropriate (e.g. defamatory) content. This basic screening allows a rapid turnaround for posting but does not provide any validation of the research methods or conclusions. Some are concerned that preprints may result in the proliferation through the internet of poor-quality research and even misinformation. Since the screening process at preprint servers does not seek to evaluate the quality of submissions, papers of varying quality may be posted. Anecdotally, however, the preprint editors at journals have so far indicated that the bioRxiv preprint papers they see are of high quality. A study that compared a group of preprints posted on bioRxiv with their associated journal article (where available) and with an equivalent group of publications in PubMed found that reporting quality at the peer-reviewed articles was higher than at the preprints (Carneiro et al., 2020). However, the results also suggest that editorial peer review has a statistically significant but small impact on improving the quality of reporting.

Public access and media coverage

Preprints are free to access and can be discovered and used by experts and the public alike – although these two groups have different backgrounds and skill sets for assessing the quality and credibility of the scientific work. The risk that a preprint will disseminate findings that will not hold up to later scrutiny does exist, and concern has been raised as to whether this could undermine public trust in scientific research. It is important to remember, however, that this is not a risk exclusive to preprints. There are numerous examples of studies published in peer-reviewed journals with public health claims (such as links between vaccination and autism) that were later debunked. The peer-review process, while providing a valuable gatekeeping framework, cannot guarantee that a study is completely free of flaws or that the conclusions in the article will hold up as new research comes to light. The stamp of approval provided by the peer-review process might in fact exacerbate the risk of misinformation if and when erroneous findings are disseminated.

In the 2020 ASAPbio stakeholder survey, the top concern of respondents was the risk of premature media coverage of preprints ('Preprint Authors Optimistic about Benefits', 2020). While media coverage of work posted on a preprint had long been the topic of discussion among stakeholders, this has come to the fore amid the COVID-19 pandemic. The distinction between a preprint and peer-reviewed clinical evidence may be clear for specialists in the relevant fields but not for non-specialized audiences or the media. To support transparency around the nature of preprints, as mentioned before, some preprint servers provide disclaimers both at the platform and on individual preprints indicating that the paper has not been certified by peer review. Organizations such as the NIH have issued tips for communicators when reporting research posted as a preprint (*Making Effective Use of Preprints*, 2020), and ASAPbio, in collaboration with a number of stakeholders, has released documents outlining guiding principles for the communication of research in the media for preprint servers, institutions, researchers and journalists ('Preprints in the Public Eye', n.d.).

Compatibility with journals

Another common concern among scientists in the early years of preprints in the life sciences was that posting a preprint might prevent them from later publishing the work in their chosen journal. While it is true that a number of journals had then precluded consideration of manuscripts previously posted as a preprint, in recent years many journals and publishers have updated their policies to take a more preprint-friendly stance. Currently, the majority of the journals in the life sciences allow or encourage preprints. For instance, among top-cited life and earth sciences journals, 91 per cent allow the posting of preprints, while the figure is about 70 per cent for the health and medical sciences (Klebel et al., 2020). It is important for researchers to check the publication policy of the journals to which they may submit prior to depositing their preprint. It is worth noting that some journals do outline restrictions concerning when the manuscript may be posted as a preprint, that is, they allow preprinting prior to or at submission to the journal but not once the paper has undergone peer review.

Intellectual property

All known preprint servers allow authors to retain copyright of their work, so from a legal perspective preprinting does not prevent authors from entering into subsequent publishing agreements ('Preprint Licensing FAQ', n.d.). However, authors must consider the license under which the preprint will be made available. The lack of clear and consistent licensing guidelines can cause concerns about the implications for future redistribution and reuse. Some preprint servers require authors to post their work under a Creative Commons CC BY license, which allows redistribution and reuse provided attribution is given to the source – this is the license that the NIH has encouraged for preprint deposition (*NIH Preprint Pilot FAQs*, n.d.). Other servers provide authors with a variety of license options. While we are not aware of any journals that refuse the submission of preprints posted under certain licenses, some hybrid journals require authors to choose an open access option when they have posted their preprint under an open license (e.g. the ASN Nutrition journals (*Author Self Archiving Policy – ASN Nutrition Journals*, n.d.)). The directory of preprint servers on the ASAPbio website provides information on the scope and practices at each preprint platform, including licensing options to help authors select a license that dovetails with their needs ('List of Preprint Servers', n.d.).

In the context of patents, authors should be aware that preprints are considered public disclosures, the same as journal articles, and thus disclosure of the work as a preprint may affect a patent application. To avoid running into such problems, researchers planning to patent their work should seek advice from their technology transfer office or relevant adviser before posting the work as a preprint.

Effectiveness of feedback

While one of the potential benefits of preprints is the possibility of having feedback on a manuscript before publication in a journal, there is currently little evidence concerning the usefulness of such feedback. The bioRxiv survey reported that 37 per cent of authors received feedback on their preprints by email and 34 per cent through in-person conversations (Sever, Roeder, et al., 2019). Such private feedback is difficult to track and quantify across the preprint ecosystem.

Some journals now allow authors to post the manuscript on a preprint server with which they are partnered; as a result, it is not uncommon for authors to post the preprint in parallel to, or even after, submission to the journal. This has led some to question whether this benefit of early feedback actually exists, and there have been calls for both preprint servers and journals to reconsider whether they would allow preprint posting if this takes place after submission to a journal (Anderson, n.d.).

Disparities in adoption

Preprint servers are free to access and present minimal editorial barriers. Therefore, they have the potential to democratize scientific communication. In the absence of financial barriers or hurdles associated with gatekeeping mechanisms, preprints allow anyone involved in research to disseminate their work independent of discipline, country or career stage. Despite this, adoption of preprints so far has been mostly driven by researchers in North America and Europe. A study of bioRxiv content concluded that countries such as the United States and the United Kingdom are overrepresented on bioRxiv relative to their overall scientific output (Abdill et al., 2020). While perhaps not a surprising trend – for example, countries with a high preprint representation overlap with those where more funders and national agencies have expressed support for preprints – the geographical disparities in preprint adoption merit further investigation to understand what the drivers and barriers are around preprints for researchers in different settings.

CHAPTER 3

SOCIAL MEDIA, COMMENTARY AND OPEN REVIEW OF PREPRINTS

Feedback on preprints can benefit authors, and where it is publicly visible, it can provide additional context for readers as well. Currently, public feedback on preprints can be found in a variety of venues and platforms.

SOCIAL MEDIA

Much of the discussion about preprints occurs on social media. A survey of bioRxiv users reported that 44 per cent of respondents received feedback via Twitter, the most common avenue by which authors receive comments (Sever, Roeder, et al., 2019).

Several factors motivate the discussion on preprints on social media. Because many researchers already use social media tools regularly, there is no barrier associated with learning a new platform, and it is also potentially more motivating to provide feedback to a known audience (e.g. Twitter followers or a Facebook group). Whether shared by the authors themselves, automated bots (like https://twitter.com/biorxivpreprint), fellow scientists or members of a more general audience, the promotion of individual preprints on social media can play a large role in their visibility and exposure. It is also correlated with other metrics of attention; for example, a high rate of tweeting of bioRxiv and medRxiv COVID-19 preprints was correlated with coverage by journalists and also a higher number of views and downloads (Fraser et al., 2021).

Social media has also played an important role in the adoption of preprints more generally. While the rate of preprints indexed monthly was less than 3 per cent of the total volume of literature appearing monthly on PubMed through

2018 (Polka & Penfold, 2020), it would be easy to come to a different conclusion by visiting certain networks on Twitter. Not all scientists use Twitter, but many of those who do are vocal advocates of preprints – as an example, Twitter conversations under the hashtag #ASAPbio in 2016 helped to create a community of users interested in preprints. The popularity of preprints among certain Twitter users is likely a confounding factor in the high rates of Twitter attention given to preprints relative to journal articles (Fraser et al., 2020; Fu & Hughey, 2019). The 'filter bubble' ('Filter Bubble', 2020) created by social media can create the perception that a behaviour or belief that is rare in the overall population is widespread. This adoption of cultural norms on Twitter was likely a driver in preprint adoption and may have influenced journal and institutional policies as well.

Facebook has so far received less attention in the context of the dissemination of scholarly work. While its impact on the visibility of preprints is lower than that of Twitter, research suggests that activity in Facebook related to science communication may have been underestimated. A study looking at Facebook activity related to publications in *PLOS ONE* in the period 2015–2017 reported that more than half of the actions (shares, reactions, etc.) took place directly between users and not via the public pages which are generally used for altmetrics studies (Enkhbayar et al., 2020). This suggests that researchers are using Facebook to share information privately and not as a channel for professional science outreach, and thus it is possible that preprints may be shared privately via Facebook to a larger scale than anticipated.

There has been growing interest in utilizing social media attention to help search or gauge interest in preprints. The web application Rxivist (Abdill & Blekhman, 2019) allows searching for bioRxiv and medRxiv preprints based on Twitter activity. Altmetric, a data science company that tracks citations and online mentions to published research (including social media as well as other sources such as the mainstream media, blogs or community forums like Reddit), tracks content in a number of preprint servers (*Repositories and Preprint Servers Tracked by Altmetric*, 2020). bioRxiv has implemented a dashboard aggregating discussion and reviews of preprints (*An Easy Access Dashboard Now Provides Links to Scientific Discussion and Evaluation of BioRxiv Preprints*, n.d.), including media/blogs and Twitter sections powered by Altmetric. Studies looking at Altmetric data for bioRxiv preprints have shown that journal articles with an associated preprint are shared more once published, across different online platforms, including Twitter, blog posts and Wikipedia (Fraser et al., 2020; Fu & Hughey, 2019).

ONLINE COMMENTARY FEATURES AT PREPRINT SERVERS

Many preprint servers allow readers to comment on individual preprints. bioRxiv and medRxiv allow readers to post comments through the Disqus platform. 14 per cent of the respondents to the bioRxiv survey reported receiving public comments via the commenting platform. Note that this figure is an overestimate for individual preprints on the platform: only 5 per cent of preprints on bioRxiv have one or more comments (Sever, Roeder, et al., 2019). On medRxiv, 9 per cent of papers received comments during its first year of operation (Krumholz et al., 2020). Several preprint servers (e.g. preprints.org, OSF preprints and Research Square) offer public commenting through built-in tools or integration with other annotation tools such as Hypothesis.

> **BOX 1. EXAMPLES OF COMMENTING OPTIONS AT PREPRINT SERVERS**
>
> **bioRxix, medRxiv**
>
> Commenting box via Disqus.
>
> **Research Square**
>
> Commenting box via own platform.
>
> **OSF Preprints**
>
> Annotation via Hypothesis.
>
> **arXiv**
>
> No commenting options.
>
> **SSRN**
>
> Commenting box via Disqus for some preprints.

THIRD-PARTY COMMENTING PROJECTS

There are a number of third-party preprint commentary projects that seek to encourage and facilitate public review of preprints.[6] The platforms and

---

[6] ASAPbio's ReimagineReview platform, developed in partnership with Wellcome and Howard Hughes Medical Institute, provides a registry of innovative peer-review projects, many of which apply to preprints (*ReimagineReview – A Registry of Platforms and Experiments Innovating around Peer Review*, n.d.).

initiatives vary in the level of detail they capture and the structure for the reviews or endorsements (see Table 1). The tool Plaudit, for example, allows users to indicate that they found a paper robust, clear or exciting with a click of a button. These endorsements are then visible to other visitors of the paper if they have the Plaudit browser extension installed, or if the publisher has integrated the tool into their site, as eLife has. Another example of structured review is Outbreak Science Rapid PREreview, which provides a multiple-choice form that allows reviewers to quickly react to papers. These reviews are then collated to provide a summary for the individual paper.

Other projects take advantage of the energy of existing communities, in many cases through journal clubs. PREreview operates live preprint journal

Table 1. Platforms for commenting and review of preprints.

Platform	Feedback collected	Eligible commenter	Anonymous commenting allowed?
Plaudit	Single-click endorsements	Anyone with ORCID	No
PREreview	Multiple-choice form and comments – via Rapid PREreviews Freeform commenting with suggested templates	Anyone with ORCID	Yes
PubPeer	Freeform commenting	Anyone who registers	Yes
preLights	Structured highlights of preprints	Contributors are selected via application process	No
Peer Community In	Traditional review	Selected reviewers	Yes
Peerage of Science	Traditional review; structured format	Selected reviewers	Yes
Review Commons	Traditional review; separate section for judging significance	Selected reviewers	Yes

clubs, replicating the experience of discussing a paper with colleagues to facilitate constructive feedback on preprints (*Live-Streamed Preprint Journal Clubs* 2020). The vision is to bring more diversity to scholarly peer review by supporting researchers (particularly those at early stages of their careers) and historically underrepresented scholars to review preprints in a constructive manner. The Sinai Immunology Review Project, organized by researchers at the Precision Immunology Institute at the Icahn School of Medicine, aims to review COVID-19 preprints to help reinforce scientific credibility (Vabret et al. 2020), in parallel with similar projects at Johns Hopkins (*2019 Novel Coronavirus Research Compendium (NCRC)*, n.d.) and Oxford University (*COVID-19 Literature Reviews – Immunology*, n.d.). The journal *Nature Reviews Immunology* has started collaborations with the Sinai Immunology Review Project and the OxImmuno Literature Initiative to publish articles reporting summaries of preprints recommended by those teams (*Watching Preprints Evolve | Nature Reviews Immunology*, 2021).

Aiming to protect those who post critical feedback and to lower the bar for participation, some platforms allow anonymous or pseudonymous commenting. PubPeer is likely the most well-known site for anonymous commenting on papers, sometimes flagging research integrity concerns that lead to corrections or even retractions of published articles; anonymity makes lodging these potentially serious accusations easier. Early career researchers, who often rely on the favour of senior colleagues for funding, favourable peer review and jobs, may be deterred from raising questions or concerns publicly about other scientists' work, even if those are relatively mild. Alternatively, they may be feeling vulnerable at the moment but willing to have comments or a review attributed to them later. Anticipating this, PREreview offers the opportunity to switch accounts from private to public, attributing peer-review activity to reviewers at a later date (*PREreview v2 Beta Debuts Today*, 2019).

Another approach to peer review of preprints is to focus on the positive. preLights, a project of the publishing organization The Company of Biologists, allows early career researchers to create posts that highlight notable or interesting preprints. These posts follow a structured format that includes a summary of the main findings, what the 'preLighter' liked about the paper and open questions (*PreLights Homepage*, n.d.). This is not unlike Faculty Opinions (formerly known as F1000 Prime) except that the latter incorporates a scoring system that tallies endorsements from multiple parties, caters to a reviewing community comprised of more senior researchers and operates by a subscription model (*Homepage – Faculty Opinions*, n.d.).

Other peer-review efforts are coordinated by editors who invite reviewers to participate in the evaluation of preprints based upon their experience in a relevant field. Peer Community In, Peerage of Science and Review Commons (operated by EMBO Press in collaboration with ASAPbio) all employ this strategy. Given that this mechanism of editor-invited review is annually responsible for coordinating millions of hours of reviewer time spent on journal-organized peer review (*Peer Review*, n.d.), there is reason to think it will be a productive strategy for preprints as well.

OVERLAY JOURNALS

The public availability of preprints gave rise to 'overlay journals' which have been described as an 'open access journal that takes submissions from the preprints deposited at an archive … and subjects them to peer-review' (*Peter Suber, 'Guide to the Open Access Movement' (Formerly: 'Guide to the FOS Movement')*, n.d.). Overlay journals coordinate an editorial assessment or peer review of publicly available papers, generally preprints. Overlay journals therefore do not produce their own article content but rather provide links to the source document.

The term 'overlay journal' was coined by Paul Ginsparg in 1996 (Ginsparg, 1997), and many overlay journals have a disciplinary focus on physics and mathematics, as overlays of the content available in arXiv. These include *Discrete Analysis*, the Episciences journals and *Advances in Combinatorics*.

The aforementioned overlay journals conduct peer review when a paper is submitted by authors. In the life sciences, several projects that refer to themselves as overlay journals have emerged with different approaches to the submission process. These include biOverlay, which closed in 2020, citing difficulties around motivating reviewers and concerns among authors that the different comments may make it more difficult for them to publish at a journal. In the case of Rapid Reviews: COVID-19 (RR:C19), the workflow is motivated by a goal to provide public expert review of papers for the sake of the community in the context of a pandemic (*Approach to Reviews – Rapid Reviews COVID-19*, n.d.). The JMIRx journals (JMIRx | Bio, JMIRx | Med, JMIRx | Psy) are described as 'superjournals', which means that 'authors no longer have to submit their manuscript', though they can still elect to do so (*What Is JMIRx?*, n.d.).

PREPRINT COMMENTARY AGGREGATORS

The availability of a variety of commenting tools and platforms that capture comments and reviews in different formats and by different contributors makes for a fragmented ecosystem in which tracking the overall activity around preprint review can be difficult. Europe PMC links out to some post-publication reviews, and a couple of platforms have been developed that aim to collect and provide visibility to preprint review activities from existing communities. Sciety is a platform developed by eLife that collects and displays preprint commentary from existing communities such as PREreview, Peer Community In and the 2019 Novel Coronavirus Research Compendium (*Sciety*, n.d.). Sciety also displays peer reviews for preprints reviewed by the journal *eLife*. Early Evidence Base is a platform developed by EMBO that aggregates and allows filtering of expert reviews on preprints posted by Review Commons, Peerage of Science, *eLife*, Peer Community In, EMBO Press and *Rapid Reviews: COVID-19* (*Accessing Early Scientific Findings | Early Evidence Base*, n.d.).

PUBLIC COMMENTARY ON PREPRINTS: CULTURE, BENEFITS, CHALLENGES

Commenting allows readers to add context, suggestions, praise and criticism where it is visible to preprint authors and other readers alike. Such comments have the potential to act as a form of moderation or quality control. For example, the infamous preprint reporting 'uncanny similarity' between SARS-CoV-2 and HIV received dozens of comments and was withdrawn by the authors a mere 48 hours after posting it (Pradhan et al., 2020). In a second example, after the user 'Preprint Now' left a comment on a paper that had originally been posted without a methods section (Habib et al., 2017), the authors responded and posted a new version correcting the oversight. In addition, some of the posted comments may actually resemble traditional review for a journal in scope and format. A study of comments on bioRxiv papers reported that 12 per cent of non-authors' comments were full review reports (Malički et al., 2021).

While commenting can have positive outcomes, concerns remain. arXiv decided not to facilitate commenting following a 2016 user survey (Rieger et al., 2016). Even those respondents who were in favour of a commenting system often added a caveat that online commenting would require a moderation system to ensure effective and collegial exchanges. Indeed, comments on bioRxiv preprints are moderated to filter content that is offensive or irrelevant. Some scientists (Laba, 2016) expressed fears that commenting, especially anonymous commenting, could foster a toxic culture, one which would disproportionately affect women, minorities and other marginalized groups. They argued that quality control is better left to the peer-review process and that authors interested in public discussions about their work can use relevant social media forums.

Some of the doubts about the value of online commenting may be linked to the results of earlier experiments in public post-publication review. One such example is PubMed Commons, which garnered around 7,500 comments in its five years of operation (Dolgin, 2018) and subsequently shuttered due to low usage. PubPeer has proven more popular with over 96,000 comments as of October 2020; nevertheless, some authors perceive the commentary posted on the site to be mostly critical as the site enables commenting by individuals who are not comfortable providing open feedback to their colleagues (Callaway, n.d.).

Another consideration is whether authors would welcome feedback on their preprint and whether their explicit consent is required to complete a review of the work. The commenting tools at preprint servers and on third-party platforms allow anyone to comment on preprints, independent of whether the authors solicited feedback or not. On the other hand, the approach by overlay journals vary, with some requiring a submission by the author and some running review based on the journal's own selection. Workflows that endorse, refer to or even reproduce openly licensed content do not (legally) require the involvement of authors. Nevertheless, cultural standards (perhaps stemming from the Inglefinger rule, discussed in a later section) dictate that each 'publication' should be unique. Therefore, if an overlay journal were to declare content 'published' without the author's consent, other journals would perceive the overlay as prior publication and be reluctant to publish the work themselves. To address these issues, both RR:C19 and JMIRx present conditional offers of publication to authors, who can then decide to accept the offer or take their manuscript to a traditional journal. As a result, there is a grey area between overlay journals, commenting platforms, review services that authors opt in to and those that, like Peer Community In, offer a 'recommendation'.

There is reason to believe that commenting on preprints could be both more prevalent and constructive than the commentary on journal articles. First, a preprint may report work that is preliminary or has not yet reached final form.

As a result, the authors can incorporate the comments into subsequent versions of the paper, to be posted to the preprint server or for submission to a journal. Commenting on preprints offers the possibility of improving the paper rather than resulting in a retraction or correction for the journal version of record, which can often carry stigma for the authors. Second, preprints have not undergone journal peer review, and as a result, public comments have a greater impact on perceived community opinion, possibly leading to a greater incentive to post.

Despite the different platforms and initiatives available for preprint commenting, the usage of these online commentary platforms is still relatively low, with only a small percentage of bioRxiv authors indicating that they had received comments on third-party review sites (Sever, Roeder, et al., 2019). As the biOverlay example shows, earlier overlay journals have not always been successful and face challenges such as finding reviewers, which is already difficult for traditional journals. In the author-independent overlay model, the possibility exists that the paper may already be under consideration at a journal, and this may act as a disincentive for potential reviewers for the overlay journal, as the reviewers may prefer to focus their efforts on the submission to a peer-reviewed journal. There is also a need to address the cultural norms around commentary and critique of scholarly work. While this has traditionally taken place in private, there is an opportunity to bring much more of this discourse into the open, and for the different stakeholders to embrace new behaviors in how they engage with both published research and public commentary on that research. In order to support a positive culture of preprint feedback, ASAPbio has convened a Working Group which is developing a set of principles for all those who engage in review and commentary of preprints (authors, reviewers and the broader community) – an initial draft of these FAST principles has been shared for community input (FAST Principles to Foster a Positive Preprint Feedback Culture', n.d.).

As the number of preprints grows, demand for curation and evaluation services that can help readers sift through a torrent of papers is likely to follow. In the context of the COVID-19 pandemic, there has been increased interest in the commentary and review of research shared via preprints. As more communities become aware of the options available, and with greater encouragement of preprint commenting by journals and funders and establishment of open commenting norms, we expect this activity will grow in the future.

CHAPTER 4
TECHNICAL INFRASTRUCTURE

Preprint servers require many of the same technical and workflow support features as repositories and journals. Therefore, it is no surprise that some of the content management platforms and tools behind existing preprint servers were originally designed for other types of content. ASAPbio's survey of the landscape of preprint platforms revealed 15 different products or services in use or in development for sharing preprints ('Surveying the Landscape of Products and Services for Sharing Preprints', 2019).

Several platforms used by preprint servers are primarily geared towards more general repositories. For instance, ePrints repository software has been used to run e-LIS and CogPrints. The Center for Open Science's Open Science Framework has been adapted to create OSF Preprints, which hosts branded versions that can be moderated by community groups (Center for Open Science, 2016). The Figshare platform, originally designed for sharing a variety of research outputs and heavily used for the deposition of datasets and figures, offers a preprint service in use for SAGE advance and TechRxiv (*Figshare Works with Preprints*, n.d.). Other servers use tools originally developed for journals: HighWire's Benchpress submission system and display tools power bioRxiv and medRxiv (Sever, Roeder, et al., 2019), and SciELO Preprints runs on the Public Knowledge Project's Open Preprint Systems, which is based on Open Journal Systems (*Open Preprint Systems | Public Knowledge Project*, n.d.). Finally, other servers use proprietary platforms, such as arXiv, SSRN and Research Square.

Regardless of the platform used, many servers require a core set of functionalities to manage submitted papers and associated metadata, facilitate the quality control and editorial processes, preserve digital content and enable discovery and access to preprints. A thorough discussion of technical infrastructure requirements and technologies is beyond the purpose of this Briefing; therefore, we will only focus on some core issues such as discovery and access, metadata and preservation.

> **BOX 2. TECHNICAL ATTRIBUTES OF SELECTED PREPRINT SERVERS**
>
> **bioRxiv, medRxiv**
> - Indexed in Google Scholar, Meta, Europe PMC, SHARE, Crossref, PubMed (NIH-supported COVID preprints only)
> - Metadata openly available via API
> - Metadata includes: Title, Identifier, Publication/deposition date, Author name(s), Abstract, Relational link to final journal publication (e.g. in Crossref metadata), Author affiliation(s), License type(s), Full-text content, References, Subject category
> - Archived in Portico
>
> **arXiv**
> - Indexed in Google Scholar, PrePubmed (q-bio only), Europe PMC, SciLit, SHARE, INSPIRE-HEP, The NASA Astrophysics Data System (ADS), The arXiv Search Interface from the National Science Library, Chinese Academy of Sciences (also in Chinese), PubMed (NIH-supported COVID preprints only)
> - Metadata openly available via API
> - Metadata includes: Title, Identifier, Publication/deposition date, Author name(s), Abstract, Relational link to final journal publication (e.g. in Crossref metadata)
> - Persistent access through mirror sites, no external preservation services used as yet
>
> **Research Square**
> - Indexed in Google Scholar, Crossref, Researcher-app, Europe PMC, PubMed (NIH-supported COVID preprints only)

- Metadata openly available via Crossref
- Metadata includes: Title, Identifier, Publication/deposition date, Author name(s), Abstract, Author affiliation(s), Funder acknowledgement(s), Subject category, Full-text content, References, License type(s), Relational link to journal publication version (where it exists)
- Archived in Portico

SSRN

- Indexed in Europe PMC, PubMed (NIH-supported COVID preprints only)
- Metadata availability: Unknown
- Metadata includes: Title, Identifier, Publication/deposition date, Author name(s), Abstract, References
- Preservation unknown

Data from asapbio.org/preprint-servers on 19 June 2021

DISCOVERY AND ACCESS

In order to be recognized as legitimate research objects by scholarly communities, preprints need to be readily discoverable. While most preprint servers offer search functionalities on their own sites, preprints must be integrated with the search tools commonly used by researchers. A number of indexing services include preprints: Google Scholar has covered preprints for many years; in the life sciences, Europe PMC began indexing preprints in 2018 (*Preprints – About – Europe PMC*, n.d.); PubMed began indexing preprints in 2020 when it announced a pilot to include preprints with NIH support relating to the SARS-CoV-2 virus and COVID-19 (*NIH Preprint Pilot*, n.d.). However, coverage of preprints in other databases is inconsistent, as Jeroen Bosman has found (*Scholarly Search Engine Comparison – Google Sheets*, n.d.). This inconsistency has led to challenges not only in discovering preprints but also in factoring them in research assessment and metrics in order to quantify the influence or impact of scholarly work.

In an ideal world, all databases and search tools would link preprints with any subsequent versions of the paper posted on the same server, on other servers or in journals and would offer viewers the option to access citations to all versions separately or in aggregate. This would not only facilitate discovery and access but also allow a more accurate and realistic understanding of the impact of a scholarly work throughout its entire life cycle. It would also remove perverse incentives that could drive journals to discourage authors from sharing preprints in the interest of optimizing their citation rankings.

METADATA

If a third party, such as an indexer, wants to make preprints discoverable, it must have access to highly structured information about the preprint. Metadata about preprints deposited by the servers to Crossref is readily accessible via Crossref's APIs (*Preprints – About – Europe PMC*, n.d.). Since 2016, Crossref has offered a preprint work type that allows the deposit of metadata highly relevant to preprints, such as the DOI of the published version of the article (Lammey, 2016). However, some servers do not use DOIs (most notably arXiv, which maintains its own persistent identification system) or use DOIs registered with DataCite, which does not offer special support for preprints. Regardless of the type of persistent identifiers used, metadata about preprints can also be provided directly by the server via an OAI-PMH endpoint (*Open Archives Initiative Protocol for Metadata Harvesting*, n.d.) or other APIs.

The metadata that is offered by these services may not be all that is collected by the preprint server, and all that is collected may not be as complete as one might expect for a journal article. This has led to calls for improvements to preprint metadata ('Four Recommendations for Improving Preprint Metadata', 2020). Preprints, which derive much of their value from the ease with which they can be posted and shared, must balance the benefits of extensive metadata with the burden this would place on depositing authors. ASAPbio, in collaboration with EMBL-EBI and Ithaka S+R, convened the #biopreprints2020 workshop to discuss such issues in January 2020. The attendees contributed to a report listing prioritized metadata (Beck et al., 2020). We placed particular focus on clarifying procedures for indicating the availability of data, on the withdrawal or removal of preprints and on managing version information in metadata. Managing versions is particularly important to ensure the accurate citation of preprints which can present a number of challenges in relation to citation format and discoverability (Hunter et al., 2020).

PRESERVATION

Digital preservation (a term used interchangeably with 'archiving') refers to a range of technical and managerial activities that support the long-term maintenance of digital content, thereby ensuring that digital objects are usable and accessible over time. Digital preservation provides an important indicator of sustainability and involves more than bitstream preservation. There is no consistent information available about arrangements between preprint platforms and third-party preservation services providers such as CLOCKSS, Internet Archive and Portico for the long-term management of digital assets. While some preprint servers report maintaining in-house backups of their files, a preservation strategy that involves a third party is likely to be more robust if the server winds down. Many preprint servers rely on Portico for the long-term preservation of their content ('List of Preprint Servers', n.d.); however, CLOCKSS also offers support for preprints (*CLOCKSS Provides 2019 Annual Update – CLOCKSS*, 2019). The Internet Archive's Scholar Search is built on full-text archiving of scholarly content, including preprints (*About Internet Archive Scholar*, n.d.).

CHAPTER 5
STAKEHOLDER PERSPECTIVES

Not surprisingly, preprints are seen differently by different stakeholders in the broader science communication space. Even within the sciences, authors, readers, funders, publishers and open science advocates all have different views (Chiarelli et al., 2019b). Authors may use preprints to share their research, readers may use them to gain access to an otherwise paywalled piece of work and librarians may want to understand best practices in the use and reuse of preprints as they seek to support faculty and contribute to the development of open access models.

AUTHORS

arXiv users have expressed a high degree of satisfaction with the server. In a survey carried out in 2016 as part of arXiv's 25th anniversary, 95 per cent of the respondents indicated they were satisfied or very satisfied (Rieger et al., 2016). Reflecting the maturity of the server, the survey had a focus on the platform's existing tools and on future potential services, rather than on motivations or potential challenges around preprint use more broadly. Interestingly, the survey results showed support for maintaining arXiv's original focus as a repository of papers, with praise for the access benefits it provides but reservations expressed about the addition of commenting tools or social media features.

As part of a survey of its users in 2019, bioRxiv asked about the motivations for posting a preprint. The main motivations were to increase awareness of the research (noted by 80 per cent), the perceived benefit to the scientific enterprise

(69 per cent), control over when the research is available (55 per cent) and to stake a priority claim (54 per cent) (Sever, Roeder, et al., 2019). In addition, when asked how having a preprint had helped them in their career, 74 per cent of respondents indicated that the preprint had increased awareness of their research. The survey also asked about any consequences of posting a preprint, and 90 per cent of respondents indicated that they had experienced no negative consequences, with only 0.7 per cent indicating they believed that the preprint had prevented them from publishing in the journal of their choice because another group published before them (Sever, Roeder, et al., 2019).

With the goal of gaining a broad set of views around the benefits and challenges of preprints, ASAPbio, in collaboration with participants in the #biopreprints2020 workshop, carried out a survey of stakeholder perspectives in the summer of 2020 ('Preprint Authors Optimistic about Benefits', 2020). The #biopreprints2020 survey received 512 responses from a broad range of stakeholders, although most responses came from researchers (369 responses). 46 per cent of researchers who responded to the survey had posted a preprint. Researchers who had authored a preprint also rated the benefits of preprints higher, and had fewer concerns about them, than those who had not posted one, suggesting that familiarity with preprinting may help mitigate some of the concerns stakeholders have. A 2020 survey among Croatian researchers on attitudes and practices around open data, open peer review and preprinting, also found that those who had posted preprints expressed a positive attitude about preprinting, although it is important to note that only a minority (11 per cent) of respondents had previously posted a preprint, and the attitude towards preprinting was neutral overall across the full sample (Baždarić et al., 2020).

A more in-depth understanding of the use and perception of preprints among early career researchers (ECRs) in comparison to more senior researchers would be beneficial. Since ECRs are often required to publish to build their careers, the delays associated with journals' peer review and publication processes can place them in a precarious situation. Preprints allow ECRs to publish their work more quickly so that they can use the preprint as proof of their productivity for funding or job applications (Sarabipour et al., 2019).

While the potential benefits of preprints may be more obvious for ECRs than for more senior researchers, in certain disciplines, some senior researchers now regularly post their latest work as a preprint prior to publication in a journal (Vale & Hyman, 2016). Some principal investigators require that every paper from their lab be posted as a preprint; others avoid posting preprints due to fear of being scooped or because they have reservations about sharing non peer-reviewed work.

There are also marked differences in preprint adoption among science disciplines. For example, the palaeontology section at bioRxiv has so far received fewer than 10 submissions per month. The lower participation from palaeontology could be due in part to the challenges around the process for naming new species, which is regulated by the International Code of Zoological Nomenclature and requires publication at a journal for the species to be recognized (Shih, 2017).

The use of preprints among chemists has also traditionally been low, mirroring their slow adoption of open access publishing, which some have claimed lagged five years after the life sciences (Carà et al., 2017). The slower preprinting trend has been linked to faster publication timelines in chemistry compared to the life sciences, which reduces the incentive for fast dissemination afforded by preprints. Another potential disincentive for chemistry researchers is that preprints are considered public disclosures and since some research in the field may have applications in industry researchers may seek to patent their discoveries. While adoption has been slower, ChemRxiv, a preprint for chemistry launched in August 2017, has experienced consistent growth and it hosts almost 20,000 preprints as of June 2021.

Discoveries in medicine can have an important impact on human health and well-being, so the dissemination of faulty medical research can be a significant risk to public health. Publishing an article in medicine tends to take longer than in the life sciences, but some medical researchers are comfortable with the slower pace of publishing because they value the stronger assessment of research via peer review more than the publication speed (Leopold et al., 2019). medRxiv was launched in 2019, six years after bioRxiv and several of the other servers for the life sciences. The adoption of preprints by one life sciences community tends to result in their uptake by communities with overlapping interests. Thus, in addition to the surge of preprints related to the COVID-19 pandemic, interdisciplinary research that touches on more clinical areas is likely to support further adoption of preprints among clinicians.

READERS FROM THE SCHOLARLY COMMUNITY

Many scholarly journals still operate under subscription models in which published articles are only accessible to those with a subscription or those who pay a fee to access the individual article. This access limitation presents challenges for independent researchers, those in disciplines or countries with lower levels of funding and those working in industry. A clear benefit of preprints from a reader's perspective is the fact that they are freely available online and accessible to anyone with an internet connection. The benefits of access that preprints afford were praised by many respondents to the 2016 arXiv survey, and 91 per cent of the respondents to the #biopreprints2020 survey rated preprints being 'free to read' as beneficial or somewhat beneficial.

THE GENERAL PUBLIC AND THE LAY MEDIA

The public availability of preprints means that they are also accessible to non-specialists such as journalists or general readers. Normally, attention to the latest peer-reviewed research tends to be restricted to the experts working in the field, with only few research findings making it to the lay press. The COVID-19 pandemic, however, has brought unprecedented attention to the latest research reports, including work posted as preprints. This public attention has raised concerns about potential misinterpretation of research findings, including how preprints could contribute to mis- or even disinformation. Some have argued that wide dissemination and media coverage of papers is best withheld until after the work is peer reviewed and published in a journal (Sheldon, 2018). Since general readers are likely unaware of the nuances of the scientific publishing process and how preprints fit into it, it's possible that preprints could be mistaken as authoritative information. A study of 457 media articles that implicitly or explicitly referred to COVID-19-related preprints found that only 57 per cent of the stories included at least one framing device to emphasize scientific uncertainty around the underlying study, that is, the story mentioned that the study was a preprint, or that it was unreviewed, preliminary and/or in need of verification. While the study points to the need for improvement and further standardization of industry practice around coverage of research posted as a preprint, the authors also note that media outlets may be more attentive to addressing scientific uncertainties when including such information is particularly relevant for public health (Fleerackers et al., 2021).

Many preprint servers include labels warning readers that the preprint is a provisional finding, has not been peer reviewed and thus should not be used to guide health-related decisions. However, some preprint servers also host peer-reviewed postprints or other research outputs, which makes the use of consistent labels across all content challenging. Furthermore, the efficacy of such labelling rests on the assumption that readers notice it and know (or are willing to learn) about the peer-review process. Such knowledge might be easier to assume if all journals were similarly transparent about the screening and evaluation processes they perform prior to publishing papers.

Media coverage of preprints is also complicated by tight journalistic timelines. Scholarly journals often provide a readily digested press release as well as a media embargo that allows journalists time to carefully consult additional experts in order to create measured coverage (Sheldon, 2018). With preprints, however, journalists have no such embargo and may experience greater pressure to rush their story out, since competitors may publish their own coverage at any time. This pressure to publish is exacerbated by another feature of preprint coverage: since preprints do not bear the presumed stamp of quality of publication by a journal, there is a greater burden on journalists to contact experts and independently verify whether a preprint is worth covering. While journalists would ideally spend *more* time reporting on a preprint, the lack of an embargo can pressure them to spend *less*.

Media coverage can be extremely influential, but it is not the only way preprints reach a wide audience. Preprints can be broadly disseminated on social media; for example, a study quantified the level of interest garnered by several preprints and showed they attracted reactions from a diverse range of non-specialist audience sectors such as mental health advocates, dog lovers, video game developers, vegans, bitcoin investors, conspiracy theorists, journalists, religious groups and political constituencies (Carlson & Harris, 2020). While this broad range of audiences and risk of misappropriation by certain groups are likely not unique to preprints, the authors of the study highlight the need for researchers to be mindful not only of how they communicate but also how they design their studies.

LIBRARIANS

Librarians have a potential role in helping researchers access the latest scholarly literature, supporting them as they make decisions on how to communicate their work and in reporting outputs for research assessment.

As the use of preprints in biomedicine has become more common, health science librarians have followed the footsteps of physical science librarians and taken steps to support faculty in making informed choices as to when and where to preprint. Some libraries have also developed resources around preprints and are carrying out training for faculty members to educate them about preprint servers, how preprints differ from journal publications and how to navigate publisher-driven workflows in relation to preprint deposition (Garrison, n.d.; Levinson, n.d.).

Librarians often support researchers' information needs by helping them identify relevant preprints and facilitating the inclusion of preprints in evidence-gathering exercises such as systematic reviews and meta-analyses. In this context, librarians have an interest in contributing to and keeping abreast of standards, practices and technologies around preprints, for example, in relation to the handling of manuscript versions, that is, preprint versions and an eventual version of record at a journal.

Many libraries are also interested in contributing to open science and supporting new business models and community-based initiatives. For instance, arXiv is partially funded by the annual membership fees paid by a group of research libraries from around the world. Library-supported initiatives such as Invest in Open Infrastructure aim to improve funding and resourcing for open technologies and systems supporting research and scholarship including preprint services (*Invest in Open Infrastructure (Page 1)*, n.d.).

FUNDERS

Funders often recognize the value of preprints because they allow faster dissemination of funded research, maximizing its reach and impact. Preprints reporting on content that might not be submitted to a journal can also help generate more published results from the same funding. Several funders (e.g. the NIH, Chan Zuckerberg Initiative or the European Research Council) have updated their policies over the last few years to include statements that encourage or even mandate the posting of preprints.

Some funders have also recognized the benefits of prompt sharing in the context of public health crises. In response to the Zika outbreak in 2016, the Wellcome Trust, along with a number of other organizations, issued a statement calling for the open sharing of information relevant to public health emergencies, including datasets and preprints (*Sharing Data during Zika and Other Global Health Emergencies | Wellcome*, 2016). More recently, in response to the

COVID-19 pandemic, the Wellcome Trust reaffirmed this position including support for the sharing of research findings via preprint servers before journal publication (*Coronavirus (COVID-19): Sharing Research Data | Wellcome* 2020).

At the end of 2020, the European Commission launched Open Research Europe (*Open Research Europe | Open Access Publishing Platform*, n.d.) a free open access platform that allows the rapid dissemination of research output from EU-funded research. The platform follows the F1000 model where submissions are posted as preprints, followed by open peer review.

PUBLISHERS

The position of publishers has shifted over time, and most journals in the life sciences now have policies compatible with preprints.

A survey carried out by Delta Think in September 2020 (O'Connell, 2020) asked respondents to rate their trust in findings presented in a preprint. Publishers reported the lowest level of trust at 45.3 per cent, whereas 59.6 per cent of researchers trusted findings in preprints. This result is not surprising given publishers' association with the peer-review process. Publishers' perspectives and engagement in the preprint domain are further elaborated in the following section.

CHAPTER 6
PREPRINTS AND PUBLISHING

Scholarly journals have traditionally fulfilled four functions in relation to scholarly work: registration (a time-stamped record of the author's work), certification (a process of evaluation of the work through peer review), dissemination (sharing the work with other peers, traditionally in the form of journals or books) and archiving to ensure the permanence of the record for future use (Guedon, 2018; Priem & Hemminger, 2012).

The use of preprints could be said to provide an alternative to several of these functions. Preprints also broaden the scholarly communication chain to allow the dissemination of earlier, more preliminary stages of the work. It is therefore not surprising that journals saw the development of the Information Exchange Groups and other related initiatives as a threat to their status and business model (Cobb, 2017).

While publishers kept the preprint phenomenon at a distance for decades, they have responded to the growth in preprint use in recent years by progressively incorporating preprints into their suite of services for researchers. The main models adopted by publishers fall into three categories:

- Acquisition of an existing preprint server: Elsevier acquired SSRN in 2016. Wiley acquired the Authorea platform in 2018 and used it to develop its Under Review service, allowing authors to deposit their manuscript as a preprint while undergoing review in parallel at a Wiley journal. Taylor & Francis acquired F1000Research in January 2020 (Schonfeld & Rieger, 2020).
- Launch of a new preprint platform: A group of chemical societies (The American Chemical Society, Royal Society of Chemistry, Chemical

Society of Japan, the Chinese Chemical Society and German Chemical Society) launched ChemRxiv in 2017. IEEE launched TechRxiv, a preprint server for electrical engineering, computer science and related areas in 2019. Cambridge University Press and SAGE have launched preprint platforms for the social sciences (Schonfeld & Rieger, 2020).

- Partnership with an existing preprint server: A major example of this approach comes from Springer Nature's partnership with Research Square's In Review service. Springer Nature is the majority stakeholder for Research Square and allows authors to post their manuscript as a preprint via the In Review platform. Launched in 2018, In Review reached the milestone of 100,000 posted preprints in August 2021 and is expected to continue its rapid expansion thanks to the inclusion of additional journals in the partnership. Other examples of partnerships between publishers and existing platforms include the many journals participating in bioRxiv and medRxiv's J2B program, which allows authors to post their paper to bioRxiv or medRxiv in parallel to the journal submission (Sever, Roeder, et al., 2019).

BOX 3. EXAMPLES OF RELATIONSHIPS BETWEEN PREPRINT SERVERS AND PUBLISHERS

bioRxiv, medRxiv
- Hosted by the academic institution Cold Spring Harbor Laboratory
- Partnerships with a range of journals

arXiv
- Hosted by the academic institution Cornell University
- Publisher independent

Research Square
- Owned by publishing services organization
- Partnership with the publisher Springer Nature

SSRN
- Owned by the publisher Elsevier

A common theme across publisher platforms is to make the transition of preprints into journal submissions – and vice versa – as smooth as possible. Although the operation and governance of the platforms may differ, publishers wish to keep researchers within their set of workflows and tools, allowing closer contact with authors at different points in the research cycle. This also provides publishers with a stepping stone to expand into services targeting earlier steps in the research process and other outputs such as data or protocols (Schonfeld & Rieger, 2020). Publishers also understand the value of preprints in accelerating the pace of scholarly communication without compromising the peer-review process and the time it requires. Coupling preprint and formal publishing processes gives publishers an opportunity to demonstrate the value provided by their editorial and peer-review processes.

Publishers' increasing involvement is likely to accelerate the use of preprints among researchers, but it also raises several questions as to what this means for the future preprint landscape. There is a potential for a substantial proportion of preprints to be available via publisher-operated platforms in an environment increasingly controlled by a few publishers. In the natural and medical sciences, the proportion of content published by the five major publishers (Elsevier, Springer Nature, Wiley-Blackwell, Taylor & Francis, American Chemical Society) grew from just over 20 per cent in 1973 to 53 per cent in 2013 (Larivière et al., 2015). It is likely that these publishers will also seek to dominate the preprint space, allowing them to keep preprint authors within their workflows. While the immediate benefit to publishers lies in the relationship to the researcher community and the networks that preprint platforms attract, publishers also have the opportunity to provide competitive tools. Publishers may view preprint servers as part of a 'razor and blades' business model where servers provides a free service that bolsters researcher satisfaction and potential engagement with other services provided by the same publisher; another potential future scenario could involve publishers introducing fees on preprints or associated services to bring financial sustainability to the preprint platforms.

The increasing presence of publishers has the potential to shift the preprint space. bioRxiv, which has played a dominant role in the adoption of preprints in the life sciences, is operated by an academic institution, and most of the current preprint servers are operated by academic communities, institutions or societies. However, the consolidation of publishing over the last decades raises the question of whether major publishers will seek to occupy an increasingly crowded space by either launching new preprint platforms or acquiring servers into their portfolios.

Preprint advocates appreciate the fact that preprints allow researchers to publish their work when they are ready. Community-based preprint servers such as arXiv and bioRxiv are publisher neutral and place no restrictions on how or when authors submit their paper to their chosen journal. Such decoupling of early sharing of research from formal publishing prevents authors from being locked into a single publisher. This freedom may be undermined if preprints are closely integrated into a publishing workflow provided by major publishers. If a few publishers consolidated ownership of preprint platforms, this would also move more of the governance and decision-making concerning server operations and business models from scholarly communities to publishers. On the other hand, closer integration with formal publishers may help with long-term sustainability of the scholarly record. A system owned by publishers might also bring more transparency and integrity to the entire publishing process from preprint to published article.

JOURNAL POLICIES

Since publishers have begun to integrate preprints into their suite of services, it is not surprising that many of them either encourage – or at least permit – preprint posting. The SHERPA/RoMEO database maintains information on publisher policies regarding the self-archiving of journal articles and lists over 1,200 publishers with policies that accept preprints (*Search - v2.Sherpa*, n.d.). This includes publishers such as Springer Nature, Elsevier or Wiley, which have adopted unified policies across their portfolios permitting preprint deposition. An informal list of academic publishers by preprint policy is also available on Wikipedia ('List of Academic Journals by Preprint Policy', 2021), and the TRANSPOSE database provides detailed journal policies around preprints (*Transpose React-App*, n.d.).

In a move underscoring the journal's support for preprints, *eLife* announced that from July 2021 it would only review manuscripts that have been already posted as a preprint (Eisen et al., 2020). Cell Press and the American Chemical Society, whose previous policies were incompatible with preprints, now consider work posted at a preprint server for publication. Not all publishers are preprint-friendly, however, and a few journals continue to reject work that has been posted as a preprint (Leopold et al., 2019).

While some journals and publishers have provided updates and clarifications around their editorial policies on preprints, a review of policies at 171 major academic journals across all disciplines found that 39 per cent still had no clear policy on whether preprints could be posted or not (Klebel et al., 2020). In their discussion document about preprints, the Committee on Publication Ethics recommended that journals develop clear policies outlining their position on preprints and that they make those publicly available (*COPE Discussion Document*, 2018).

The need to adapt to the increasing presence of preprints has presented challenges for some editors. Some have suggested that preprints should be considered a 'publication' since they share many of the characteristics of journal publications: preprints have content that resembles that of a journal article, and many preprint servers assign DOIs and are included in indexing services. Peer-reviewed journals generally will not consider work that has been published previously. This restrictive policy originates from the so-called 'Ingelfinger rule' implemented in 1969 by Franz Ingelfinger, then editor-in-chief of *The New England Journal of Medicine (NEJM)*. The policy stipulated that the journal would not publish work already published elsewhere, in either other media or other journals ('Ingelfinger Rule', 2019). The policy sought to protect the originality of publications in the *NEJM* and to prevent the duplicate ('redundant') publication of work that could lead to bias in the literature. Some editors have interpreted the rule broadly and considered a preprint publication as a breach of the Ingelfinger rule. Yet, as preprint adoption has increased, many editors have revised or reversed their position; the *NEJM* itself, in an editorial in relation to research on the COVID-19 outbreak, encouraged authors to share their work as a preprint (Rubin et al., 2020).

PEER REVIEW

Peer review after publication

The certification of scholarly work via peer review is one of the core functions provided by scholarly journals. In its traditional form, peer review at journals takes place behind closed doors, with editors privately approaching two or three reviewers to evaluate the work and provide a recommendation on publication; upon a positive evaluation, the journal then publishes the article. Peer review thus happens *before* publication, and the article can then undergo post-publication peer review by the community, either via private conversations or through public commenting channels provided by the journal or other platforms.

Dissemination of the paper via a preprint allows the publication of the work without peer review and thus results in a different sequence where publication happens first, and peer review takes place *after* publication. The peer review of the manuscript can then take place via a traditional process involving two or three reviewers (either through services that provide peer review of preprints or through submission to a journal) or via post-publication commenting by members of the community. Given that the paper is publicly available, review by selected reviewers and by community members can take place at the same time.

Peer review: the tension between speed and quality

When considering the traditional peer-review framework that takes place before publication, speed is often mentioned as one of the main downsides. The peer-review process at journals takes weeks, if not months or even years, while preprints allow work to be shared in a matter of days. During the COVID-19 pandemic, the urgency to share work relevant to the public health crisis resulted not only in a surge in the posting of preprints but also in an acceleration of the peer-review process at journals. An analysis of 669 publications at medical journals showed that turnaround times decreased by 47 days on average for COVID-19-related articles (Horbach, 2020). This is a welcome response from journals, but it is unlikely that such acceleration of the review process can be sustained over time or at a larger scale: the same study reported no change in turnaround times for non-COVID-19 publications.

While some say that preprints help relieve the time pressure on the peer-review process, others have questioned the need for quicker publication. Will a focus on speed compromise the quality of the peer-review process? Is it better to have information available more quickly or information which is published slowly but more rigorously evaluated?

The peer-review process provides a certification that published findings have passed scrutiny by experts. It provides a quality-control mechanism, but it is by no means a perfect process. Peer review generally involves assessment by an editor and two or three reviewers; as the number of publications has increased and scientific papers have become more complex, it has become increasingly difficult to assign reviewers who can cover all the potential techniques and analyses reported in a single manuscript. Preprints, on the other hand, make work available to a broad community of potential experts who, in principle, can scrutinize all components of the manuscript, though reader attention on preprints is much more diffuse than during the journal peer-review process.

In the context of the COVID-19 pandemic, the peer-review process for several high-profile articles reporting results on the use of hydroxychloroquine for patients with COVID-19 came under scrutiny after readers raised concerns about the studies. The fast review turnaround for an article in *International Journal of Antimicrobial Agents* (Gautret et al., 2020) raised some eyebrows; the paper subsequently received an Expression of Concern. Two publications in the *New England Journal of Medicine* (Mehra, Desai, et al., 2020) and *The Lancet* (Mehra, Ruschitzka, et al., 2020) were retracted just weeks after publication following concerns about the veracity of the claims and lack of access to the underlying data. The peer-review process for these articles had been accelerated due to COVID-19, which suggested to some that speed may have come at the expense of quality. Weeks after the retraction, *The Lancet* announced changes to its editorial policies, including the involvement of at least one reviewer with dataset expertise for studies based on large, real-world datasets (The Editors of the Lancet Group, 2020).

The debate around the balance between speed and quality in the peer-review process is not a new one, but the COVID-19 pandemic has exacerbated this tension and brought it to the foreground; it is likely that the debate will persist. While some believe that we should capitalize on the accelerated peer review we observed in 2020 and rapidly review all papers, others instead advocate for a more measured process that allows an in-depth evaluation to take place. A slower and more thorough review process can increase the rigor of published articles and reduce the number of post-publication corrections and retractions – as well as increase the reproducibility of published work. Others suggest that preprints can offer a 'release valve', reducing pressure on journals to make compromises in the name of accelerating their peer-review process, while still allowing findings to be publicly accessible in provisional form (Anderson, 2017).

REPRODUCIBILITY

Concerns about a 'reproducibility crisis' in research are not new (Baker, 2016). Studies attempting to replicate previously published work have reported different degrees of success (Nosek & Errington, 2017), and the number of retractions, while remaining a very small proportion of the published literature, has substantially increased over the last decade.[7] The 'publish or perish' culture is often linked to perverse incentives that favour publication in prestigious journals over best practices such as thoroughness of the design, full reporting or sharing of underlying data or materials. Concerns about reproducibility predate the use of preprints in the life sciences, and there are different views on whether preprints have a positive or negative effect on reproducibility. Critics have raised concerns as to whether preprints may result in the widespread sharing of underdeveloped papers, lowering the quality and reproducibility of available work. They also note that the peer-review process plays an important role by allowing an evaluation of the level of reproducibility of the work before it reaches publication.

While peer review can certainly play a role in ensuring higher reproducibility of published work, the notion that peer review assures reproducibility places a considerable burden on the reviewers to evaluate and reproduce the research after the work is already completed. A different position on reproducibility places a stronger focus earlier in the research process, encouraging better research practices starting from the inception of the work, to the running of experiments and data collection, to the writing of the paper.

A number of initiatives have arisen over the last decade that encourage the sharing of research outputs at earlier stages of the research process and beyond the format of a traditional journal article. OSF introduced pre-registration, which allows researchers to post their research plan including design and analysis outlines. A pre-registration can be posted publicly, to allow community scrutiny, or privately, to provide a dated record of the stipulated analysis plans as a means to prevent potential bias in the analysis upon data collection. OSF now hosts over 300,000 pre-registrations. Registered Reports are a publication format that allows researchers to submit their research plan for peer review, prior to the start of the study; if the plan passes peer review, the journals that consider Registered Reports commit to publishing the eventual article once the research is completed, independent on whether the results are positive or negative, and thus provide an avenue to prevent publication bias. Some journals will publish both the research plan and the final article reporting the results as separate

---

[7] PubMed lists just over 1,100 retractions published in the period 2000–2009; this number increased fivefold in the following decade with 5,834 retractions published in 2010–2019.

articles; some others only publish the latter. After a timid start, support for the model is growing and now almost 300 journals have implemented Registered Reports (*Registered Reports*, n.d.).

Another innovation relates to the publication of protocols. While the publication of study protocols for clinical research was supported by a number of medical journals, publication avenues for laboratory protocols were rare. To address this gap, protocols.io was created in 2014 as a platform for the open sharing of protocols. protocols.io now hosts almost 8,000 protocols and has entered a partnership with PLOS to allow the protocols posted on the platform to also be published in *PLOS ONE* ('Submit Your Lab and Study Protocols to PLOS ONE!', 2021).

The type of content that can be posted at preprint servers varies, and some platforms (e.g. bioRxiv or arXiv) only allow the posting of full research manuscripts; however, a number of other preprint servers (e.g. OSF Preprints, AfricArXiv) allow the posting of any scholarly content and thus the sharing of Registered Reports, protocols and other preliminary outputs such as posters at any stage of the research cycle. Researchers can also link other outputs (data, protocol, etc.) when they share their work via a preprint server; preprints therefore serve as a means and a vehicle for increasing the reproducibility of research at earlier and at multiple stages of the research process (Puebla, 2021).

By receiving feedback from other community members on a preliminary version of the work and associated data, researchers could improve their papers and in turn increase the reproducibility of an eventual journal publication. Efforts such as Registered Reports and support for data sharing open up the research process much earlier than a journal publication, and as their adoption in the life sciences increases, preprints are likely to play a role in such reproducibility efforts.

CHAPTER 7
PREPRINTS AND RESEARCHER ASSESSMENT

Researchers' publication choices are driven in large part by their scholarly communities and how funding agencies and institutions recognize and reward their work. Some institutions explicitly reward publication only in certain journals, meaning that sharing work through preprints and other 'grey literature' will not be directly helpful to a researcher's career (McKiernan et al., 2019). Over the last years, there have been calls to make research assessment more transparent and to move away from relying on single journal-based metrics for the assessment of research productivity and impact. The 2013 San Francisco Declaration on Research Assessment (DORA) decries the use of 'journal-based metrics, such as Journal Impact Factors, as a surrogate measure of the quality of individual research articles, to assess an individual scientist's contributions, or in hiring, promotion, or funding decisions'. DORA recommends that funding agencies and institutions consider the value of all outputs in addition to journal publications. The DORA declaration has been signed by over 2,000 organizations and 17,000 individuals ('Signers', n.d.). In the United Kingdom, a review of the role of metrics in research assessment and management produced the Metric Tide report in 2015, outlining recommendations for a responsible use of metrics and calling for the adoption of a variety of metrics to provide a wider view on the quality and impact of research (Wilsdon et al., 2015). The same year, the Leiden Manifesto outlined 10 principles for research assessment, including recommendations for the use of qualitative assessment in addition to quantitative metrics and the need to recognize field specificities in publication and citation practices as part of research evaluation frameworks (Hicks et al., 2015). More recently, in early 2020, the Chinese government announced reforms of the research and higher education evaluation system to reduce the reliance on the Science Citation Index and journal impact factors, which had played a prominent role in research evaluation processes in the country in recent years (Tao, 2020).

The recognition that research should be evaluated on its own merits based on different metrics (such as article-level metrics and others), rather than using the journal as a proxy of quality, has opened the possibility for preprints to be considered in research assessment frameworks as evidence of productivity, separately from the publication venue.

FUNDER POLICIES

As discussed in an earlier section, the number of funders that allow preprints to be cited to demonstrate progress for grant applications and reports has substantially increased over the last five years. It should however be noted that funders' positions on preprints vary. Some have not released formal preprint policies, like the US National Science Foundation (NSF), which funds work in physics and mathematics – disciplines with long traditions of preprint adoption – although there is anecdotal evidence that preprints are cited in NSF progress reports. The Australian Research Council (ARC) previously implemented a restrictive policy which did not allow the citation of preprints. This funder's decision to disqualify a number of fellowship applications which cited preprints met strong criticism by the research community (Australian Research Council under Pressure after Funding Rule Angers Academic Community, 2021) and the ARC subsequently revised its policy to permit the citation of preprints (Australian Research Council, 2021).

In 2016, the Simons Foundation was the first funder to institute a policy encouraging researchers to post preprints. Since then, several philanthropic and public funding agencies have allowed preprints to be included in grant applications and reports (see Table 2). Arguably the most influential development for U.S. biomedical researchers was the 2017 announcement by NIH, the world's largest public biomedical funder, encouraging the use of preprints (NOT-OD-17–050 guide notice). Most of those funders that have expressed support for preprints encourage, but do not require, the use of preprints. Notable examples of funders that now mandate preprint deposition include the Chan Zuckerberg Initiative and Aligning Science Against Parkinson's.

Table 2. Funding agencies announcements in support of preprints as proof of research productivity ('Funder Policies', n.d.).

May 2016	Simons Foundation Autism Research Initiative (SFARI) encourages posting of preprints, in parallel with or before submission to a peer-reviewed journal.
September 2016	Helmsley Trust encourages prospective and existing grantees to list preprints in their applications and interim reports.
December 2016	Human Frontiers Science Program announces that applicants may list preprints in proposals and interim and final reports.
January 2017	Wellcome Trust permits citation of preprints in grant applications and end-of-grant reports. The Medical Research Council, United Kingdom, allows preprints to be cited in grant and fellowship applications if the preprint is less than five years old at the time of application. The Howard Hughes Medical institute recognizes preprints as evidence of productivity.
March 2017	The National Institutes of Health encourages researchers to cite preprints as proof of productivity.
May 2017	Cancer Research UK allows (and encourages) deposition of preprints and preprint citation in funding applications.
June 2017	The Biotechnology and Biological Sciences Research Council encourages grantees to share their pre-peer-review manuscripts via preprint servers.
September 2017	The Canadian Institutes of Health Research notes recognition of preprints as an 'important vehicle for the dissemination of research results'.
October 2017	Le Centre National de la Recherche Scientifique (CNRS, France) states that preprints should be taken into account in the processes of hiring, evaluation and promotion of researchers as well as project evaluation.
March 2018	Chan Zuckerberg Initiative requires deposition of preprints.
August 2018	The European Research Council announces that it will accept preprints as evidence of research work in grant applications.
Fall 2018	The Doris Duke Charitable Foundation includes preprints as evidence of contributions to research, both in grant applications and in progress reports.
September 2019	The Serrapilheira Institute (Brazil) recommends that researchers deposit articles as preprints 'before or upon submission'.

October 2019	Aligning Science Across Parkinson's requires that publications related to funded work must be submitted to a preprint server before or concurrent to the first submission to a journal.
March 2020	Michael J. Fox Foundation for Parkinson's Research announces that articles resulting from its funding must be posted in an open access preprint repository.
June 2020	L'Agence Nationale de la Research (France) indicates that preprints will be accepted as part of applications in their call for proposals within the Plan d'action 2020.

This support by several funding agencies is likely to have influenced preprint adoption in biomedicine over the last few years. In the 2019 survey of bioRxiv users, 42 per cent of respondents noted the ability to cite the research in a grant application as one of their motivations for posting work on bioRxiv (Sever, Roeder, et al., 2019). In the #biopreprints2020 survey, 70 per cent of respondents who self-identified as researchers considered the possibility to demonstrate progress in the context of evaluation for grants or job applications as a highly or somewhat beneficial aspect of preprints ('Preprint Authors Optimistic about Benefits', 2020).

NATIONAL AND INSTITUTIONAL POLICIES

Preprints have also been included in some national frameworks, which in turn influence university policy. The French national research alliances for the environment (AllEnvi) and Life Sciences and Health (AvieSan) released a statement endorsing the use of preprints for the evaluation of both projects and individual researchers in the context of hiring and promotion (*Les preprints sont une forme recevable de communication scientifique*, n.d.). In the United Kingdom, preprints are valid research outputs for submission to the Research Excellence Framework (REF), a national assessment of the research conducted at UK universities carried out by the higher education funding bodies (*Preprints Are Valid Research Outputs for REF2021 –ASAPbio*, 2019). REF scores inform the allocation of around £2 billion/year of national funding for research, so REF is a major driver of UK institutional policy and researcher behaviour.

An additional important element of research assessment relates to hiring and promotion processes. While the processes at individual universities and research institutions vary, both in approach and on the level of public information available (Fernandes et al., 2020), there are examples of institutions that have stated support for the use of preprints as evidence of productivity. The University of California Davis (United States) has announced to faculty the addition of a 'preprints' category in the online faculty achievements database, and the dean of UFRGS Research (Federal University of Rio Grande do Sul, Brazil) has encouraged the use of preprints and their inclusion in projects, work plans and activity reports. In addition, job postings for some positions at U.S. universities have made mention of preprints (summarized at the ASAPbio website ('University Policies and Statements on Hiring, Promotion, and Journal License Negotiation', n.d.)).

Preprints can also influence hiring decisions in the absence of formal policies. In a survey of job applications among early career researchers (270 applicants, with a majority of U.S. representation), most of whom (85 per cent) were working in the life sciences, 55 per cent of respondents had posted at least one preprint and 40 per cent had an active preprint not yet published in a journal at the time of the faculty job application (Fernandes et al., 2020). Several respondents mentioned that preprints were helpful in their job search as proof of productivity before journal publication. The same survey included a limited sample of 15 U.S. faculty members on faculty search committees, two-thirds of whom reported viewing preprints listed in candidate applications favourably.

CHAPTER 8
PREPRINTS AS AN OPEN SCIENCE TOOL

Preprints align quite closely with the open science ethos of collaboration and broad dissemination of research works. The European Commission defines open science as an 'approach to the scientific process based on cooperative work and new ways of diffusing knowledge by using digital technologies and new collaborative tools' (Commission, n.d.). Preprints are one of the elements within a broader open science ecosystem that includes digital notebooks, the deposition of protocols, datasets and code and open access journal publishing.

Traditional journal publication typically accompanies a full and complete disclosure of all methods and reagents necessary to reproduce the science. However, minimal screening on preprints (and the fact that norms for data sharing with preprints are still developing) creates situations in which preprints may represent incomplete disclosures. For example, some preprints have been posted without a methods section, prompting critiques that they resemble 'ads' rather than scientific papers. However, communities are pushing to develop their own norms and expectations around data deposition with preprints (#ASAPpdb: Structural biologists commit to releasing data with preprints, n.d.).

One of the goals of open science is to enable reuse of research findings, both by other researchers and by machines. However, most preprint servers make papers available only in PDF format, which limits reliable text mining and automated content extraction, two of the potential uses for open science. While some servers provide HTML and XML versions of the paper, the conversion to those formats can be an expensive process.

ARE PREPRINTS OPEN ACCESS?

Where do preprints fit within open access initiatives? At the beginning of the open access movement, many thought that self-archiving (green open access) would enable widespread access without representing a huge threat to publishers. This did not materialize, however, and the growth in open access has mostly been driven by publication in gold open access journals (*Rate of Growth for CC BY Articles in Fully-OA Journals Continues for OASPA Members*, 2019). The number of open access articles has consistently grown since 2000, and this trend is likely to continue with the support of mandates by funders and national bodies as well as the Plan S initiative, which came into effect in January 2021.

Preprints allow authors to deposit a copy of their work in a publicly available format and thus align to some of the principles of open access. The original definition of open access, however, requires both public access and the permission to reuse content. Thus, whether a preprint equates to an open access version of the work depends on the license under which it is posted and whether the work is equivalent to that which is finally published in a journal.

> **BOX 4. EXAMPLES OF LICENSING OPTIONS AT SELECTED PREPRINT SERVERS**
>
> **bioRxiv, medRxiv**
>
> A choice of licenses: CC0, CC BY, CC BY-NC, CC BY-NC-ND, no reuse; with no preference for which license chosen
>
> **arXiv**
>
> A choice of licenses: CC0, CC BY, CC BY-SA 4.0, CC BY-NC-SA 4.0, non-exclusive license to distribute, any other CC license; with no preference for which license chosen
>
> **Research Square**
>
> Authors must use CC BY license
>
> **OSF Preprints**
>
> A choice of licenses: CC0, CC BY, no license; with no preference for which license chosen

At a minimum, preprint servers must obtain permission from the authors to display and distribute their content on their site; these are essentially the terms of arXiv's standard license. At the other end of the spectrum, some preprint servers require that authors post their paper under a CC BY license, a Creative Commons license which allows reproduction and reuse of the material if attribution is given to the original authors. Some servers allow authors to choose from a suite of licenses, including Creative Commons licenses that restrict usage for commercial purposes (CC BY-NC), the ability to make derivative works (CC BY-ND), or stipulate that all derivatives must carry a similar license (CC BY-SA) ('Preprint Licensing FAQ', n.d.).

When submitting to preprint servers that provide a range of license options, authors tend to make conservative choices. For example, most preprints on arXiv are licensed under the arXiv standard license, not one of the Creative Commons licenses. An analysis of bioRxiv and medRxiv preprints in the context of the COVID-19 pandemic showed that no license (i.e. no permission granted for reuse) and CC BY-NC-ND remained the top license choices (Fraser et al., 2021). These trends are not necessarily surprising as licensing is often a confusing subject for researchers. In the #biopreprints2020 survey, 56 per cent of the respondents scored 'Uncertainty about copyright and licensing of preprints' as concerning or very concerning ('Preprint Authors Optimistic about Benefits', 2020). There is also some legacy perception that choosing a license that allows reuse of the preprint may create challenges for the eventual publication of the paper at a journal. While some publishers used to only consider papers posted as a preprint that did not have Creative Commons licenses (e.g. PNAS, FASEB, IOP (McKenzie, 2017)), many publishers have now revised their policies. In general, authors should be able to post their preprint under the license of their choice and publish subsequent versions of the paper under a different license or even assign the copyright to a publisher for journal publication ('Preprint Licensing FAQ', n.d.).

The free availability of preprints has also prompted discussions around whether they can satisfy open access mandates by funders and institutions. The proponents of the 'Plan U' proposal have argued that a universal funder mandate requiring that grantees post their manuscripts first as a preprint would be easier to implement than open access policies governing the version of record, since the latter requires a shift in business models for many journals (Sever, Eisen, et al., 2019). The feasibility of this proposal rests on whether preprints are satisfactory substitutes for the version of an article ultimately published in a journal after peer review. An analysis of this question must take into account two factors: first, whether the journal and server policies restrict which version of the preprint (or indeed, postprint) can be shared and, second, in the presence of such policies, whether peer review meaningfully improves manuscripts.

Regarding the first point, while many servers outside of the biomedical sciences allow the deposition of postprints, some preprint servers in biomedicine (e.g. bioRxiv) do not allow authors to post manuscripts accepted for publication. Furthermore, some journals don't allow authors to incorporate into preprint versions the changes made as part of the journal's peer review process. As a result, compliance with these policies means that only the initial submission to the journal can be posted at the preprint server. This link between the initial journal submission and the preprint record is further reinforced by the close integration between several preprint servers (e.g. bioRxiv or Research Square) and journals' manuscript submission systems.

Second, do manuscripts change meaningfully through the peer review process? Some studies have found little difference between preprints in arXiv and journal versions (Klein et al., 2019) and similarities between reporting quality of preprints in bioRxiv and journal publications (Carneiro et al., 2020). An analysis of preprints posted in the initial months of 2020 which were subsequently published in journals found that the number of figures changed little between preprint and published articles, and the conclusions reported in abstracts remained the same for a majority of papers; however, the conclusions of 6 per cent of non-COVID-19-related and 15 per cent of COVID-19-related abstracts did undergo a discrete change by the time of publication (Polka et al., 2021).

Overall, preprint server and journal policies that prohibit updates to preprints jeopardize their ability to serve as substitutes for open access versions of record (green open access). Instead, these policies mean that some researchers are only able to access the preprint version, which is similar but may have important differences compared to the version later published in a journal. This situation may compromise the ability of preprints to mitigate existing challenges around access to the literature in settings where researchers cannot afford journal subscriptions and need to rely on preprints as an accessible although unvetted version of the paper.

MATERIALIZING THE FULL POTENTIAL OF PREPRINTS AS AN OPEN SCIENCE TOOL

The potential for preprints to advance open science could improve if manuscripts were shared earlier in the research process. The earlier sharing of ongoing work would allow more open and earlier feedback, bringing transparency to the scientific process and opportunities for collaboration within the community.

Preprints offer much greater flexibility than journals in terms of the length and format of papers that can be posted. Earlier sharing of ongoing research via preprints would reap the benefits of visibility, feedback and credit for the authors, and the possibility to add new results and refine the interpretation of the findings as the research progresses, via the posting of subsequent versions. Preprints can also be used to share work with the community that is currently not included in journal submissions: negative or inconclusive results, replications or refutations of published work, troubleshooting of protocols and reports of the development of discrete reagents or tools. As the types of research outputs continue to diversify, new and additional uses of preprints will emerge.

Utilizing preprints to share research findings as the work progresses would allow a more fluid dissemination that aligns to the research cycle and is thus less dependent on the traditional journal publication framework and timelines. Researchers can disseminate their ongoing work well before journal submission, rather like presenting posters at conferences, with the additional advantage that the preprint provides a time-stamped record that is searchable and discoverable. Earlier sharing of work via preprints would provide a powerful tool to encourage the broader collaborative approach that is one of the pillars of open science.

CHAPTER 9
CONCLUSION AND OPEN QUESTIONS

The landscape of preprints in the sciences, particularly in life sciences, has changed dramatically over the last five years. The use of preprints in these disciplines has increased steadily, including an explosion during the COVID-19 crisis. We don't know what the dynamics of preprint use will be after the pandemic. Will the use of preprints in COVID-19 catalyse further adoption or will the trend revert to a slower growth? We are optimistic that the adoption of preprints will continue, driven by greater familiarity with preprints across different research communities and the increased integration with publishers.

At this stage of early adoption, there needs to be more emphasis on how preprints are being used and supported by scholars in developing countries. As practices and initiatives develop, tool builders, funders and servers should ensure that they provide a level playing field for different research communities and avoid recreating some of the hierarchical structures that have characterized the traditional journal publication process. We should encourage social practices and infrastructure that bring diversity and equity into preprint adoption.

The COVID-19 pandemic has further strengthened the uptake of preprints due to the urgency of the grand challenge faced. Both preprint and journal article submissions increased during 2020, putting additional pressure on scientists involved in the peer-review process or the screening and moderation of papers submitted to preprint servers. Although there is great value in this broader adoption of preprints, there are also concerns that the flow of preprints will add to ongoing concerns about 'information overload': will preprints add to researchers' struggle to stay abreast of new developments? And does the need to potentially check through multiple versions of a paper add to the complexity in shifting through the latest reports? As scientific outputs continue to grow rapidly, researchers are faced with the challenge of trying to identify credible research. This increasing workload also further strains the peer-review system since both scientific journals and preprint servers rely on the participation of researchers. arXiv maintains a baseline network of more than 140 moderators (scientists with expertise in different related subject domains) to assist with the processing of almost 600 papers submitted each day. If submissions continue to grow, there will be a need to develop automated tools to help screen preprints so that this burden does not become overwhelming to the volunteer researchers. We are seeing early steps in this space with the use of SciScore and other automated tools to screen and provide automated reports on medRxiv and bioRxiv COVID-19 preprints (Weissgerber et al., 2021).

As preprint adoption grows and we build evidence on whether preprints generate wider impact for research and potentially help reduce waste in funding, we may see additional funders and institutions update their policies and assessment frameworks to incorporate preprints. Research evaluation frameworks can take time to evolve, but as we have seen through DORA and other initiatives, a number of researchers and institutions are taking steps towards more inclusive evaluation systems that account for a broader range of research outputs (Hatch & Curry, 2020). It will be interesting to monitor the evolution of research assessment frameworks to see if additional funders and institutions around the world adopt preprint-supportive policies and whether there is increased adoption of preprints in those settings.

One of the main areas of debate in coming years is likely to relate to the longer-term financial sustainability of preprint servers. If the adoption of preprints in the life sciences continues to grow, and with this the number of preprint platforms, sustainability is likely to become an increasingly pressing topic. One of the concerns about preprints is their ability to secure the steady resources (technologies, expertise, policies, visions, standards, etc.) required to maintain and enhance the value of a service based on a user community's needs (Rieger, 2012). Preprints emerged as a 'public good', and preprint platforms provide a free service to both authors and readers; at the same time, many of the existing preprint services lack a scalable and transparent business model. Unlike publishers and societies that generate revenues through different business models such as subscription, article-processing charges or membership fees, there is not yet a similar financial model for community-based preprints, which rely mostly on grants or gifts from foundations. Some servers accept donations – a model that supports other public platforms such as Wikipedia – and could also explore agreements with institutions that would contribute membership fees to sustain the server (as already in place at arXiv). However, we cannot single out preprints in this grand challenge. Some argue that there is already enough money in the system. U.S. academic libraries, for instance, spend about $7 billion a year on resources, but how would we know where to start in redistributing these funds to ensure the sustainability of new and open forms of scientific communication?

There are some emerging publication models that promise to be more financially durable and scalable. For instance, ChemRxiv is built on a partnership involving the world's five largest chemical science societies. As more publishers adopt preprints, it is possible they will also seek either to offset the costs or even generate a profit. An early example of this is Research Square's offer for fee-based statistical and methodology checks on preprints posted on their In Review platform. While funding bodies have supported the founding of several of the current preprint platforms, they have not always supplied ongoing funding for infrastructure development and maintenance activities. As preprints become more integrated into the scientific communication flows, funders might consider supporting both the design of innovative and transformative features and the daily maintenance essential for preprints to serve as reliable and trusted services. Ultimately, the sustainability of preprints will depend on service providers' ability to establish inclusive and transparent governance systems and diverse revenue streams.

The increase in the number of journals triggered the evolution of rankings such as the journal impact factor and the development of publishing industry standards. Those standards sought to support good research practice, to guide researchers in their publication choices and to differentiate publishing venues, ranging from the excellent to the 'predatory'. There are considerably fewer preprint servers than there are journals (dozens of preprint servers vs. thousands of journals). At the same time, many preprint servers have emerged in just a few years, and as the number of servers grows, some seek to establish standards for what constitutes a 'reputable' preprint server. As initial steps, for instance, Europe PMC and PubMed have created their own criteria for preprint servers (*NIH Preprint Pilot*, n.d.; *Preprints – About – Europe PMC*, n.d.). The coming years may see the development of more standards in relation to issues such as screening practices, licensing and preservation. Such standards can guide researchers as they consider posting their work to preprint servers and also create a framework of expectations for newcomers into the ecosystem.

As the scholarly communication ecosystem evolves and broadens with the addition of new formats such as preprints, there are growing concerns that this proliferation of formats may make it harder to keep track of different versions and identify the version of record. Our concept of what constitutes the scholarly record is broadening as we understand better how ideas evolve from initial research design to data gathering, analysis and the sharing of early results. There is an increasing emphasis on sharing outputs from the stage of the initial research design all the way to the peer-reviewed publication. One of the challenges in this broadening knowledge ecosystem is interconnecting the different nodes in which preprints, peer-reviewed articles, conference papers and presentations, supporting data and code, as well as related comments and amendments, can be discovered and interpreted by researchers. This is not a trivial task as scholarly communication involves a complex sociotechnical infrastructure composed of technologies, standards, policies, workflows and practices that require time to adjust and adapt.

Such an adjustment involves aligning the activities of many stakeholders including researchers, publishers, technology providers, standards developers and funders. Prominent publishers are increasingly interested in expanding their workflows to incorporate preprints into the publication cycle. From the perspective of publishers, incorporating preprints into the process could accelerate the pace of scholarly communication without compromising the peer-review process. More importantly, publishers may be able to introduce a more efficient and consistent layer of quality control than has been available through some existing preprint services. On the other hand, such a close alignment with the journal-based publishing process may interfere with the publisher-agnostic and community-based forums that disciplinary communities have created.

There is also interest in revising the system of peer review and curation. A variety of platforms and initiatives have emerged in recent years seeking to incentivize engagement with preprints and to capture reader reactions, comments and reviews. We are likely to see further innovation and experimentation with review formats in the coming years. The circumstances around research dissemination in the context of the COVID-19 pandemic have reminded us of the need to scrutinize all research reports, independently of whether they are posted as a preprint, a journal article or in another format. The pandemic has also highlighted that, despite heightened interest around scientific work, the public is largely unfamiliar with the iterative and collaborative nature of the scientific process. Science requires multiple steps of repetition, validation and scrutiny before establishing new scientific evidence. In this context, we are obliged to be more transparent about the nature of the scientific process and the stages of dissemination and validation of scientific discoveries.

Transparency and reproducibility are at the centre of open science and are increasingly being recognized as essential indicators of research quality and credibility. Therefore, one of the factors that will determine the future of preprints is the degree to which existing services support access to underlying data and analysis. However, this is easier said than done as research data management is still evolving and will require significant investment of both resources and expertise. Therefore, preprint services should approach this area cautiously and creatively and explore collaboration opportunities with existing data services and repositories.

Preprints are an increasingly important tool within the broader and ever-changing ecosystem of research communication. In addition to broadening access to research outputs and enabling innovation in research assessment, they also promise to expand participation in research and its communication. Because preprints enable community feedback and discussion at a point when it can actually influence the trajectory of a project – rather than after it is already fixed in a publisher's version of record – they can strengthen individual studies as well as scientific networks. As research becomes more open, preprints are likely to become a necessary pillar in realizing the full potential for a more diverse and collaborative research environment.

REFERENCES

2016 Meeting. (n.d.). *ASAPbio*. Retrieved 5 November 2020, from https://asapbio.org/meeting-information.

2019 Novel Coronavirus Research Compendium (NCRC). (n.d.). 2019 Novel Coronavirus Research Compendium (NCRC). Retrieved 21 January 2021, from https://ncrc.jhsph.edu/.

Abdill, R. J., Adamowicz, E. M., & Blekhman, R. (2020). International authorship and collaboration across bioRxiv preprints. *ELife*, *9*, e58496. https://doi.org/10.7554/eLife.58496.

Abdill, R. J., & Blekhman, R. (2019). Tracking the popularity and outcomes of all bioRxiv preprints. *ELife*, *8*, e45133. https://doi.org/10.7554/eLife.45133.

About Internet Archive Scholar. (n.d.). Retrieved 6 November 2020, from https://scholar-qa.archive.org/about#howitworks.

Accessing early scientific findings | Early Evidence Base. (n.d.). Retrieved 5 July 2021, from https://eeb.embo.org/refereed-preprints/review-commons.

African Scientists Launch Their Own Preprint – Scientific American. (n.d.). Retrieved 20 November 2020, from https://www.scientificamerican.com/article/african-scientists-launch-their-own-preprint/.

An easy access dashboard now provides links to scientific discussion and evaluation of bioRxiv preprints. (n.d.). Retrieved 2 July 2021, from https://connect.biorxiv.org/news/2021/05/14/dashboard.

Anderson, K. (n.d.). *Should preprint policies be revised?* Retrieved 5 November 2020, from https://thegeyser.substack.com/p/are-preprint-policies-upside-down.

Anderson, K. (2017, March 28). *The tincture of time – should journals return to slower publishing practices?* The Scholarly Kitchen. https://scholarlykitchen.sspnet.org/2017/03/28/the-tincture-of-time-should-journals-return-to-slower-publishing-practices/.

Approach to Reviews – Rapid Reviews COVID-19. (n.d.). Rapid Reviews COVID-19. Retrieved 20 November 2020, from https://rapidreviewscovid19.mitpress.mit.edu/reviewapproach.

#ASAPpdb: Structural Biologists Commit to Releasing Data with Preprints." (n.d.) ASAPbio. Retrieved 21 January 2021, from https://asapbio.org/asappdb.

Australian Research Council. *"Adjustments to the ARC's Position on Preprints.* (September 13, 2021). Australian Research Council. https://www.arc.gov.au/news-publications/media/communiques/adjustments-arcs-position-preprints.

REFERENCES

Australian Research Council under Pressure after Funding Rule Angers Academic Community. (2021, August 23). *The Guardian.* http://www.theguardian.com/education/2021/aug/24/australian-research-council-under-pressure-after-funding-rule-angers-academic-community.

Author Self Archiving Policy – ASN Nutrition Journals. (n.d.). Oxford Academic. Retrieved 20 November 2020, from https://academic.oup.com/journals/pages/self_archiving_policy_asn.

Baker, M. (2016). 1,500 scientists lift the lid on reproducibility. *Nature News, 533*(7604), 452. https://doi.org/10.1038/533452a.

Barrett, S. C. H. (2018). Proceedings B 2017: The year in review. *Proceedings of the Royal Society B: Biological Sciences, 285*(1870), 20172553. https://doi.org/10.1098/rspb.2017.2553.

Barsh, G. S., Bergman, C. M., Brown, C. D., Singh, N. D., & Copenhaver, G. P. (2016). Bringing PLOS Genetics editors to preprint servers. *PLOS Genetics, 12*(12), e1006448. https://doi.org/10.1371/journal.pgen.1006448.

Baumann, A., & Wohlrabe, K. (2020). Where have all the working papers gone? Evidence from four major economics working paper series. *Scientometrics, 124*(3), 2433–2441. https://doi.org/10.1007/s11192-020-03570-x.

Baždarić, K., Vrkić, I., Arh, E., Mavrinac, M., Marković, M. G., Bilić-Zulle, L., Stojanovski, J., & Malicki, M. (2020). Attitudes and practices of open data, preprinting, and peer-review – a cross sectional study on Croatian scientists. *BioRxiv*, 2020.11.25.395376. https://doi.org/10.1101/2020.11.25.395376

Beck, J., Ferguson, C. A., Funk, K., Hanson, B., Harrison, M., Ide-Smith, M., Lammey, R., Levchenko, M., Mendonça, A., Parkin, M., Penfold, N., Pfeiffer, N., Polka, J., Puebla, I., Rieger, O. Y., Rittman, M., Sever, R., & Swaminathan, S. (2020). *Building trust in preprints: Recommendations for servers and other stakeholders.* OSF Preprints. https://doi.org/10.31219/osf.io/8dn4w.

Berg, J. M., Bhalla, N., Bourne, P. E., Chalfie, M., Drubin, D. G., Fraser, J. S., Greider, C. W., Hendricks, M., Jones, C., Kiley, R., King, S., Kirschner, M. W., Krumholz, H. M., Lehmann, R., Leptin, M., Pulverer, B., Rosenzweig, B., Spiro, J. E., Stebbins, M., . . . Wolberger, C. (2016). Preprints for the life sciences. *Science, 352*(6288), 899–901. https://doi.org/10.1126/science.aaf9133.

Binfield, P., & Hoyt, J. (2013, April 3). *Who killed the prePrint, and could it make a return?* Scientific American Blog Network. https://blogs.scientificamerican.com/guest-blog/who-killed-the-preprint-and-could-it-make-a-return/.

BioRxiv Reporting. (n.d.). Retrieved 6 November 2020, from https://api.biorxiv.org/reports/category_summary.

Callaway, E. (n.d.). Pioneer behind controversial PubPeer site reveals his identity. *Nature News.* https://doi.org/10.1038/nature.2015.18261.

Carà, P. D., Ciriminna, R., & Pagliaro, M. (2017). Has the time come for preprints in chemistry? *ACS Omega, 2*(11), 7923–7928. https://doi.org/10.1021/acsomega.7b01190.

Carlson, J., & Harris, K. (2020). Quantifying and contextualizing the impact of bioRxiv preprints through automated social media audience segmentation. *PLOS Biology, 18*(9), e3000860. https://doi.org/10.1371/journal.pbio.3000860.

Carneiro, C. F. D., Queiroz, V. G. S., Moulin, T. C., Carvalho, C. A. M., Haas, C. B., Rayêe, D., Henshall, D. E., De-Souza, E. A., Amorim, F. E., Boos, F. Z., Guercio, G. D., Costa, I. R., Hajdu, K. L., van Egmond, L., Modrák, M., Tan, P. B., Abdill, R. J., Burgess, S. J., Guerra, S. F. S., . . . Amaral, O. B. (2020). Comparing quality of reporting between preprints and peer-reviewed articles in the biomedical literature. *Research Integrity and Peer Review, 5*(1), 16. https://doi.org/10.1186/s41073-020-00101-3.

REFERENCES 67

Center for Open Science. (2016, December 5). *The Center for Open Science releases a comprehensive, open source preprints solution.* https://www.cos.io/about/news/center-open-science-releases-comprehensive-open-source-preprints-solution.

Chiarelli, A., Johnson, R., Pinfield, S., & Richens, E. (2019a). *Accelerating scholarly communication: The transformative role of preprints.* Zenodo. https://doi.org/10.5281/ZENODO.3357727.

Chiarelli, A., Johnson, R., Pinfield, S., & Richens, E. (2019b). Preprints and scholarly communication: An exploratory qualitative study of adoption, practices, drivers and barriers. *F1000Research, 8,* 971. https://doi.org/10.12688/f1000research.19619.2.

CLOCKSS Provides 2019 Annual Update – CLOCKSS. (2019, August 1). https://clockss.org/2019/08/clockss-provides-2019-annual-update/.

Cobb, M. (2017). The prehistory of biology preprints: A forgotten experiment from the 1960s. *PLOS Biology, 15*(11), e2003995. https://doi.org/10.1371/journal.pbio.2003995.

Commission. (n.d.). *What is open science? Introduction.* FOSTER FACILITATE OPEN SCIENCE TRAINING FOR EUROPEAN RESEARCH. Retrieved 16 November 2020, from https://www.fosteropenscience.eu/content/what-open-science-introduction.

COPE Discussion Document: Preprints. (2018). Committee on Publication Ethics. https://doi.org/10.24318/R4WByao2.

Coronavirus action plan: A guide to what you can expect across the UK. (n.d.). GOV.UK. Retrieved 6 November 2020, from https://www.gov.uk/government/publications/coronavirus-action-plan/coronavirus-action-plan-a-guide-to-what-you-can-expect-across-the-uk.

Coronavirus (COVID-19): Sharing research data | Wellcome. (2020, January 31). https://wellcome.org/coronavirus-covid-19/open-data.

COVID-19 Literature Reviews – Immunology. (n.d.). Retrieved 21 January 2021, from https://www.immunology.ox.ac.uk/covid-19/covid-19-immunology-literature-reviews.

COVID-19 Portfolio | Home. (n.d.). Retrieved 6 November 2020, from https://icite.od.nih.gov/covid19/search/.

Dimensions. (n.d.). Retrieved 20 November 2020, from https://app.dimensions.ai/discover/publication.

Dolgin, E. (2018). PubMed Commons closes its doors to comments. *Nature.* https://doi.org/10.1038/d41586-018-01591-4.

Eisen, M. B., Akhmanova, A., Behrens, T. E., Harper, M. H., Weigel, D., & Zaidi, M. (2020). Peer review: implementing a "publish, then review" model of publishing. *eLife, 9*:e64910. https://doi.org/10.7554/eLife.64910

Electronic Colloquium on Computational Complexity. (2019). In *Wikipedia.* https://en.wikipedia.org/w/index.php?title=Electronic_Colloquium_on_Computational_Complexity&oldid=923179879.

Elsevier. (2016, May 17). *SSRN – the leading social science and humanities repository and online community – Joins Elsevier.* Elsevier Connect. https://www.elsevier.com/connect/ssrn-the-leading-social-science-and-humanities-repository-and-online-community-joins-elsevier.

Enkhbayar, A., Haustein, S., Barata, G., & Alperin, J. P. (2020). How much research shared on Facebook happens outside of public pages and groups? A comparison of public and private online activity around PLOS ONE papers. *ArXiv:1909.01476 [Cs].* http://arxiv.org/abs/1909.01476.

FAST Principles to Foster a Positive Preprint Feedback Culture. (n.d.). ASAPbio. Retrieved 10 September 2021, from https://asapbio.org/fast-principles.

Fernandes, J. D., Sarabipour, S., Smith, C. T., Niemi, N. M., Jadavji, N. M., Kozik, A. J., Holehouse, A. S., Pejaver, V., Symmons, O., Bisson Filho, A. W., & Haage, A. (2020). A survey-based analysis of the academic job market. *ELife*, *9*, e54097. https://doi.org/10.7554/eLife.54097.

Figshare works with Preprints. (n.d.). Retrieved 6 November 2020, from https://knowledge.figshare.com/type-of-client/preprints.

Filter bubble. (2020). In *Wikipedia*. https://en.wikipedia.org/w/index.php?title=Filter_bubble&oldid=986099321.

Fleerackers, A., Riedlinger, M., Moorhead, L., Ahmed, R., & Alperin, J. P. (2021). Communicating scientific uncertainty in an age of COVID-19: an investigation into the use of preprints by digital media outlets. *Health Communication*, *0*(0), 1–13. https://doi:10.1080/10410236.2020.1864892.

Four recommendations for improving preprint metadata. (2020, April 8). *Scholarly Communications Lab | ScholCommLab*. https://www.scholcommlab.ca/2020/04/08/preprint-recommendations/.

Fraser, N., Brierley, L., Dey, G., Polka, J. K., Pálfy, M., Nanni, F., & Coates, J. A. (2021). The evolving role of preprints in the dissemination of COVID-19 research and their impact on the science communication landscape. *PLOS Biology*, *19*(4), e3000959. https://doi.org/10.1371/journal.pbio.3000959.

Fraser, N., Momeni, F., Mayr, P., & Peters, I. (2020). The relationship between bioRxiv preprints, citations and altmetrics. *Quantitative Science Studies*, *1*(2), 618–638. https://doi.org/10.1162/qss_a_00043.

Fu, D. Y., & Hughey, J. J. (2019). Releasing a preprint is associated with more attention and citations for the peer-reviewed article. *ELife*, *8*, e52646. https://doi.org/10.7554/eLife.52646.

Funder policies. (n.d.). *ASAPbio*. Retrieved 5 November 2020, from https://asapbio.org/funder-policies.

Garisto, D. (n.d.). *Preprints make inroads outside of physics*. Retrieved 5 November 2020, from http://www.aps.org/publications/apsnews/201909/preprints.cfm.

Garrison, J. (n.d.). *Library guides: Preprints: A how-to guide: Home*. Retrieved 6 November 2020, from https://instr.iastate.libguides.com/c.php?g=49365&p=318314.

Gautret, P., Lagier, J.-C., Parola, P., Hoang, V. T., Meddeb, L., Mailhe, M., Doudier, B., Courjon, J., Giordanengo, V., Vieira, V. E., Tissot Dupont, H., Honoré, S., Colson, P., Chabrière, E., La Scola, B., Rolain, J.-M., Brouqui, P., & Raoult, D. (2020). Hydroxychloroquine and azithromycin as a treatment of COVID-19: Results of an open-label non-randomized clinical trial. *International Journal of Antimicrobial Agents*, *56*(1), 105949. https://doi.org/10.1016/j.ijantimicag.2020.105949.

Ginsparg, P. (1997). Winners and losers in the Global Research Village. *The Serials Librarian*, *30*(3–4), 83–95. https://doi.org/10.1300/J123v30n03_13.

Ginsparg, P. (2016). Preprint déjà vu. *The EMBO Journal*, *35*(24), 2620–2625. https://doi.org/10.15252/embj.201695531.

Guedon, J.-C. (2018, May 17). *Unleashing knowledge with open access*. COAR 2018 Annual Meeting and General Assembly (COAR2018), Hamburg, Germany. https://doi.org/10.5281/zenodo.1250185.

Habib, N., Basu, A., Avraham-Davidi, I., Burks, T., Choudhury, S. R., Aguet, F., Gelfand, E., Ardlie, K., Weitz, D. A., Rozenblatt-Rosen, O., Zhang, F., & Regev, A. (2017). DroNc-Seq: Deciphering cell types in human archived brain tissues by massively-parallel single nucleus RNA-seq. *BioRxiv*, 115196. https://doi.org/10.1101/115196.

Hatch, A., & Curry, S. (2020). Changing how we evaluate research is difficult, but not impossible. *ELife*, *9*, e58654. https://doi.org/10.7554/eLife.58654.

Hicks, D., Wouters, P., Waltman, L., de Rijcke, S., & Rafols, I. (2015). Bibliometrics: The Leiden Manifesto for research metrics. *Nature News*, *520*(7548), 429. https://doi.org/10.1038/520429a.

Homepage – Faculty Opinions. (n.d.). Retrieved 5 July 2021, from https://facultyopinions.com/.

Horbach, S. P. J. M. (2020). Pandemic publishing: Medical journals strongly speed up their publication process for COVID-19. *Quantitative Science Studies*, *1*(3), 1056–1067. https://doi.org/10.1162/qss_a_00076.

Hoyt, J. (2019, September 3). *PeerJ Preprints to stop accepting new preprints Sep 30th 2019 –PeerJ Blog*. https://peerj.com/blog/post/115284881747/peerj-preprints-to-stop-accepting-new-preprints-sep-30-2019/.

Hunter, I. H., Kleshchevich, I., & Rosenblum, B. (2020, September 18). *Guest post – what's wrong with preprint citations?* The Scholarly Kitchen. https://scholarlykitchen.sspnet.org/2020/09/18/guest-post-whats-wrong-with-preprint-citations/.

Ingelfinger rule. (2019). In *Wikipedia*. https://en.wikipedia.org/w/index.php?title=Ingelfinger_rule&oldid=918775768.

Invest in Open Infrastructure (Page 1). (n.d.). Invest in Open Infrastructure. Retrieved 6 November 2020, from https://investinopen.org/about/.

Johansson, M. A., Reich, N. G., Meyers, L. A., & Lipsitch, M. (2018). Preprints: An underutilized mechanism to accelerate outbreak science. *PLOS Medicine*, *15*(4), e1002549. https://doi.org/10.1371/journal.pmed.1002549.

Kaiser, J. (2014, November 11). *BioRxiv at 1 year: A promising start.* Science | AAAS. https://www.sciencemag.org/news/2014/11/biorxiv-1-year-promising-start.

Klebel, T., Reichmann, S., Polka, J., McDowell, G., Penfold, N., Hindle, S., & Ross-Hellauer, T. (2020). Peer review and preprint policies are unclear at most major journals. *PLOS ONE*, *15*(10), e0239518. https://doi.org/10.1371/journal.pone.0239518.

Klein, M., Broadwell, P., Farb, S. E., & Grappone, T. (2019). Comparing published scientific journal articles to their pre-print versions. *International Journal on Digital Libraries*, *20*(4), 335–350. https://doi.org/10.1007/s00799-018-0234-1.

Krichel, Thomas. (1997, February). *About NetEc, with special Reference to WoPEc.* http://openlib.org/home/krichel/hisn.html.

Krumholz, H. M., Bloom, T., Sever, R., Rawlinson, C., Inglis, J. R., & Ross, J. S. (2020). Submissions and downloads of preprints in the first year of medRxiv. *JAMA*, *324*(18), 1903–1905. https://doi.org/10.1001/jama.2020.17529.

Kwon, D. (2020). How swamped preprint servers are blocking bad coronavirus research. *Nature*, *581*(7807), 130–131. https://doi.org/10.1038/d41586-020-01394-6.

Laba, I. (2016, April 10). ArXiv, comments, and 'quality control'. *The Accidental Mathematician.* https://ilaba.wordpress.com/2016/04/10/arxiv-comments-and-quality-control/.

Lammey, R. (2016, November 2). *Preprints are go at Crossref!* [Website]. Crossref. https://www.crossref.org/blog/preprints-are-go-at-crossref/.

Larivière, V., Haustein, S., & Mongeon, P. (2015). The oligopoly of academic publishers in the digital era. *PLOS ONE*, *10*(6), e0127502. https://doi.org/10.1371/journal.pone.0127502

Leopold, S. S., Haddad, F. S., Sandell, L. J., & Swiontkowski, M. (2019). Clinical Orthopaedics and Related Research, The Bone & Joint Journal, the Journal of Orthopaedic Research, and The Journal of Bone and Joint Surgery will not accept clinical research manuscripts previously posted to preprint servers*. *JBJS*, *101*(1), 1–4. https://doi.org/10.2106/JBJS.18.01215.

Les preprints sont une forme recevable de communication scientifique. (n.d.). Archives. Retrieved 20 November 2020, from http://archives.cnrs.fr/insb/article/888.

Levinson, C. (n.d.). *Levy Library Guides: Preprints: The Basics: Home.* Icahn School of Medicine at Mount Sinai. Retrieved 6 November 2020, from https://libguides.mssm.edu/preprints/home.

List of academic publishers by preprint policy. (2021). In *Wikipedia.* https://en.wikipedia.org/wiki/List_of_academic_publishers_by_preprint_policy.

List of preprint servers: Policies and practices across platforms. (n. d.). *ASAPbio.* Retrieved 2 July 2021, from https://asapbio. org/preprint- servers.

Live-Streamed Preprint Journal Clubs: We can help! (2020, April 2). PREreview Blog. https://content.prereview.org/live-streamed-preprint-journal-clubs-we-can-help/.

Mahase, E. (2020). Covid-19: Demand for dexamethasone surges as RECOVERY trial publishes preprint. *BMJ, 369.* https://doi.org/10.1136/bmj.m2512.

Making Effective Use of Preprints. (2020, August 18). National Institutes of Health (NIH). https://www.nih.gov/about-nih/what-we-do/science-health-public-trust/perspectives/science-health-public-trust/science-health-public-trust/making-effective-use-preprints-tips-communicators.

Malički, M., Jeromčić, A., ter Riet, G., Bouter, L. M., Ioannidis, J. P. A., Goodman, S. N., & Aalbersberg, Ij. J. (2020, November 10). *Preprint servers' policies, submission requirements, and transparency in reporting and research integrity recommendations | Medical Journals and Publishing | JAMA | JAMA Network.* https://jamanetwork.com/journals/jama/fullarticle/2772748.

Malički, M., Malički, M., Costello, J., Alperin, J. P., Alperin, J. P., & Maggio, L. A. (2021). Analysis of single comments left for bioRxiv preprints till September 2019. *Biochemia Medica, 31*(2), 0–0. https://doi.org/10.11613/BM.2021.020201.

Mathematical Physics Preprint Archive. (n.d.). Retrieved 30 June 2021, from https://psrc.aapt.org/items/detail.cfm?ID=127.

McKenzie, L. (2017). Biologists debate how to license preprints. *Nature News.* https://doi.org/10.1038/nature.2017.22161.

McKiernan, E. C., Schimanski, L. A., Muñoz Nieves, C., Matthias, L., Niles, M. T., & Alperin, J. P. (2019). Use of the Journal Impact Factor in academic review, promotion, and tenure evaluations. *ELife, 8,* e47338. https://doi.org/10.7554/eLife.47338.

MDPI. (2016, July 28). Introducing preprints: A multidisciplinary open access preprint platform. *MDPI Blog.* https://blog.mdpi.com/2016/07/28/introducing-preprints-a-multidisciplinary-open-access-preprint-platform/.

Mehra, M. R., Desai, S. S., Kuy, S., Henry, T. D., & Patel, A. N. (2020). Retraction: Cardiovascular disease, drug therapy, and mortality in Covid-19. *New England Journal of Medicine, 382*(26), 2582–2582. https://doi.org/10.1056/NEJMc2021225.

Mehra, M. R., Ruschitzka, F., & Patel, A. N. (2020). Retraction – hydroxychloroquine or chloroquine with or without a macrolide for treatment of COVID-19: A multinational registry analysis. *The Lancet, 395*(10240), 1820. https://doi.org/10.1016/S0140-6736(20)31324-6.

NIH Preprint Pilot. (n.d.). Retrieved 6 November 2020, from https://www.ncbi.nlm.nih.gov/pmc/about/nihpreprints/.

NIH Preprint Pilot FAQs. (n.d.). Retrieved 6 November 2020, from https://www.ncbi.nlm.nih.gov/pmc/about/nihpreprints-faq/.

Nosek, B. (2017, April 19). *Public goods infrastructure for preprints and innovation in scholarly communication.* https://www.cos.io/blog/public-goods-infrastructure-preprints-and-innovation-scholarly-communication.

Nosek, B. A., & Errington, T. M. (2017). Making sense of replications. *ELife*, *6*, e23383. https://doi.org/10.7554/eLife.23383.

Nouri, S. N., Cohen, Y. A., Madhavan, M. V., Slomka, P. J., Iskandrian, A. E., & Einstein, A. J. (2020). Preprint manuscripts and servers in the era of coronavirus disease 2019. *Journal of Evaluation in Clinical Practice 27*, 16–21. https://doi.org/10.1111/jep.13498.

O'Connell, A. (2020, September 22). *News & views: Preprints and COVID-19: Findings from our PRW survey*. Delta Think. https://deltathink.com/news-views-preprints-and-covid-19-findings-from-our-prw-survey/.

Open Archives Initiative Protocol for Metadata Harvesting. (n.d.). Retrieved 6 November 2020, from https://www.openarchives.org/pmh/.

Open Preprint Systems | Public Knowledge Project. (n.d.). Retrieved 6 November 2020, from https://pkp.sfu.ca/ops/.

Open Research Europe | Open Access Publishing Platform. (n.d.). Retrieved 2 July 2021, from https://open-research-europe.ec.europa.eu/.

Pandemic Preprints – A Duty of Responsible Stewardship. (2021, April 27). The BMJ. https://blogs.bmj.com/bmj/2021/04/27/pandemic-preprints-a-duty-of-responsible-stewardship/.

Peer Review: How We Found 15 Million Hours of Lost Time | AJE. (n.d.). Retrieved 5 November 2020, from https://www.aje.com/arc/peer-review-process-15-million-hours-lost-time/.

Penfold, N. C., & Polka, J. K. (2020). Technical and social issues influencing the adoption of preprints in the life sciences. *PLOS Genetics*, *16*(4), e1008565. https://doi.org/10.1371/journal.pgen.1008565.

Peter Suber, 'Guide to the Open Access Movement' (formerly: 'Guide to the FOS Movement'). (n.d.). Retrieved 6 November 2020, from http://legacy.earlham.edu/~peters/fos/guide.htm.

Petrou, C. (2020, August 10). *Guest post – MDPI's remarkable growth*. The Scholarly Kitchen. https://scholarlykitchen.sspnet.org/2020/08/10/guest-post-mdpis-remarkable-growth/.

Polka, J. K., Dey, G., Pálfy, M., Nanni, F., Brierley, L., Fraser, N., & Coates, J. A. (2021). Preprints in motion: Tracking changes between posting and journal publication. *BioRxiv*, 2021.02.20.432090. https://doi.org/10.1101/2021.02.20.432090.

Polka, J. K., & Penfold, N. C. (2020). *Biomedical preprints per month, by source and as a fraction of total literature* [Data set]. Zenodo. https://doi.org/10.5281/zenodo.3819276.

Pradhan, P., Pandey, A. K., Mishra, A., Gupta, P., Tripathi, P. K., Menon, M. B., Gomes, J., Vivekanandan, P., & Kundu, B. (2020). Uncanny similarity of unique inserts in the 2019-nCoV spike protein to HIV-1 gp120 and Gag. *BioRxiv*, 2020.01.30.927871. https://doi.org/10.1101/2020.01.30.927871.

PreLights Homepage. (n.d.). PreLights. Retrieved 6 November 2020, from https://prelights.biologists.com/.

Preprint. (2020). In *Wikipedia*. https://en.wikipedia.org/w/index.php?title=Preprint&oldid=989164213.

Preprint authors optimistic about benefits: Preliminary results from the #bioPreprints2020 survey. (2020, July 27). ASAPbio. https://asapbio.org/biopreprints2020-survey-initial-results.

Preprint FAQ. (n.d.). ASAPbio. Retrieved 5 November 2020, from https://asapbio.org/preprint-info/preprint-faq.

Preprint licensing FAQ. (n.d.). ASAPbio. Retrieved 7 November 2020, from https://asapbio.org/licensing-faq.

Preprints are valid research outputs for REF2021 – ASAPbio. (2019, August 8). https://asapbio.org/preprints-valid-for-ref2021.

Preprints in the Public Eye. (n.d.). ASAPbio. Retrieved 5 November 2020, from https://asapbio.org/preprints-in-the-public-eye.

Preprints – About – Europe PMC. (n.d.). Retrieved 6 November 2020, from https://europepmc.org/Preprints.

PREreview v2 beta debuts today. (2019, September 20). PREreview Blog. https://content.prereview.org/prereview-v2/.

Priem, J., & Hemminger, B. H. (2012). Decoupling the scholarly journal. *Frontiers in Computational Neuroscience, 6*, 19. https://doi.org/10.3389/fncom.2012.00019.

Puebla, I. (2021). Preprints: A tool and a vehicle towards greater reproducibility in the life sciences. *Journal for Reproducibility in Neuroscience, 2.* https://doi.org/10.31885/jrn.2.2021.1465.

Pulverer, B. (2016). Preparing for preprints. *The EMBO Journal, 35*(24), 2617–2619. https://doi.org/10.15252/embj.201670030.

Rate of growth for CC BY articles in fully-OA journals continues for OASPA members. (2019, July 3). OASPA. https://oaspa.org/growth-continues-for-oaspa-member-oa-content/.

Registered Reports. (n.d.). Retrieved 2 July 2021, from https://www.cos.io/initiatives/registered-reports.

ReimagineReview – A registry of platforms and experiments innovating around peer review. (n.d.). Retrieved 6 November 2020, from https://reimaginereview.asapbio.org/.

Repositories and preprint servers tracked by Altmetric. (2020, October 6). Digital Science. https://help.altmetric.com/support/solutions/articles/6000242541-repositories-and-preprint-servers-tracked-by-altmetric.

Rieger, O. Y. (2012). Sustainability: Scholarly repository as an enterprise. *Bulletin of the American Society for Information Science and Technology, 39*(1), 27–31. https://doi.org/10.1002/bult.2012.1720390110.

Rieger, O. Y., Steinhart, G., & Cooper, D. (2016). arXiv@25: Key findings of a user survey. 23. https://arxiv.org/abs/1607.08212.

Rieger, Oya Y. (2020). Preprints in the spotlight. *Ithaka S+R.* https://sr.ithaka.org/publications/preprints-in-the-spotlight/

Roger Schonfeld and Oya Rieger. (2020, May 27). *Publishers invest in preprints.* The Scholarly Kitchen. https://scholarlykitchen.sspnet.org/2020/05/27/publishers-invest-in-preprints/.

Rubin, E. J., Baden, L. R., Morrissey, S., & Campion, E. W. (2020). Medical journals and the 2019-nCoV outbreak. *New England Journal of Medicine, 382*(9), 866–866. https://doi.org/10.1056/NEJMe2001329.

Sarabipour, S., Debat, H. J., Emmott, E., Burgess, S. J., Schwessinger, B., & Hensel, Z. (2019). On the value of preprints: An early career researcher perspective. *PLOS Biology, 17*(2), e3000151. https://doi.org/10.1371/journal.pbio.3000151.

Scholarly search engine comparison – Google Sheets. (n.d.). Retrieved 6 November 2020, from https://docs.google.com/spreadsheets/d/1ZiC-UuKNse8dwHRFAyhFsZsl6kG0Fkgaj5gttdwdVZEM/edit#gid=1016151070.

Sciety. (n.d.). Sciety. Retrieved 5 July 2021, from https://sciety.org/.

Search – V2.sherpa. (n.d.). Retrieved 6 November 2020, from https://v2.sherpa.ac.uk/romeo/search.html.

Serghiou, S., & Ioannidis, J. P. A. (2018). Altmetric scores, citations, and publication of studies posted as preprints. *JAMA, 319*(4), 402. https://doi.org/10.1001/jama.2017.21168.

Sever, R., Eisen, M., & Inglis, J. (2019). Plan U: Universal access to scientific and medical research via funder preprint mandates. *PLOS Biology, 17*(6), e3000273. https://doi.org/10.1371/journal.pbio.3000273.

Sever, R., Roeder, T., Hindle, S., Sussman, L., Black, K.-J., Argentine, J., Manos, W., & Inglis, J. R. (2019). bioRxiv: The preprint server for biology. *BioRxiv,* 833400. https://doi.org/10.1101/833400.

Severin, A., Egger, M., Eve, M. P., & Hürlimann, D. (2020). Discipline-specific open access publishing practices and barriers to change: An evidence-based review. *F1000Research*, *7*, 1925. https://doi.org/10.12688/f1000research.17328.2.

Sharing data during Zika and other global health emergencies | Wellcome. (2016, February 10). https://wellcome.org/news/sharing-data-during-zika-and-other-global-health-emergencies.

Sheldon, T. (2018). Preprints could promote confusion and distortion. *Nature*, *559*(7715), 445. https://doi.org/10.1038/d41586-018-05789-4.

Shih, I. (2017, August 23). *Encouraging palaeontologists to stop hiding the bones.* https://www.natureindex.com/news-blog/encouraging-palaeontologists-to-stop-hiding-the-bones.

Signers. (n.d.). *DORA.* Retrieved 5 July 2021, from https://sfdora.org/signers/.

Soderberg, C. K., Errington, T. M., & Nosek, B. A. (2020). Credibility of preprints: An interdisciplinary survey of researchers. *Royal Society Open Science*, *7*(10), 201520. https://doi.org/10.1098/rsos.201520.

SSRN launches a new network dedicated to biology – BioRN. (2017, June 7). https://www.elsevier.com/about/press-releases/science-and-technology/ssrn-launches-a-new-network-dedicated-to-biology-biorn.

Submit your Lab and Study Protocols to PLOS ONE! (2021, February 9). *The Official PLOS Blog.* https://theplosblog.plos.org/2021/02/submit-your-lab-and-study-protocols-plos-one/.

Surveying the landscape of products and services for sharing preprints. (2019, February 13). *ASAPbio.* https://asapbio.org/preprint-products.

Tao, T. (2020, February 27). *New Chinese policy could reshape global STM publishing.* The Scholarly Kitchen. https://scholarlykitchen.sspnet.org/2020/02/27/new-chinese-policy-could-reshape-global-stm-publishing/.

The Editors of the Lancet Group. (2020). Learning from a retraction. *The Lancet*, *396*(10257), 1056. https://doi.org/10.1016/S0140-6736(20)31958-9.

Till, James E. (2001). *Predecessors of preprint servers.* https://arxiv.org/html/physics/0102004.

Transpose React-app. (n.d.). Retrieved 6 November 2020, from https://transpose-publishing.github.io/#/.

Univekiity, G. W. (n.d.). *Informal communication among scientists: A study of the Information Exchange Group Program.* 68. https://web.archive.org/web/20170212101029/http://www.dtic.mil/dtic/tr/fulltext/u2/726650.pdf.

University policies and statements on hiring, promotion, and journal license negotiation. (n.d.). *ASAPbio.* Retrieved 7 November 2020, from https://asapbio.org/university-policies.

Vabret, N., Samstein, R., Fernandez, N., & Merad, M. (2020). Advancing scientific knowledge in times of pandemics. *Nature Reviews Immunology*, *20*(6), 338. https://doi.org/10.1038/s41577-020-0319-0.

Vale, R. D. (2015). Accelerating scientific publication in biology. *Proceedings of the National Academy of Sciences*, *112*(44), 13439–13446. https://doi.org/10.1073/pnas.1511912112.

Vale, R. D., & Hyman, A. A. (2016). Priority of discovery in the life sciences. *ELife*, *5*, e16931. https://doi.org/10.7554/eLife.16931.

Watching preprints evolve | Nature Reviews Immunology. (2021). https://www.nature.com/articles/s41577-020-00489-5.

Weissgerber, T., Riedel, N., Kilicoglu, H., Labbé, C., Eckmann, P., ter Riet, G., Byrne, J., Cabanac, G., Capes-Davis, A., Favier, B., Saladi, S., Grabitz, P., Bannach-Brown, A., Schulz, R., McCann, S., Bernard, R., & Bandrowski, A. (2021). Automated screening of COVID-19 preprints: Can we help authors to improve transparency and reproducibility? *Nature Medicine*, *27*(1), 6–7. https://doi.org/10.1038/s41591-020-01203-7.

What is JMIRx? (n.d.). JMIR Publications. Retrieved 20 November 2020, from https://support.jmir.org/hc/en-us/articles/360034752692-What-is-JMIRx-.

REFERENCES

Wilsdon, J., Allen, L., Belfiore, E., Campbell, P., Curry, S., Hill, S., Jones, R., Kain, R., Kerridge, S., Thelwall, M., Tinkler, J., Viney, I., Wouters, P., Hill, J., & Johnson, B. (2015). *The metric tide: Report of the independent review of the role of metrics in research assessment and management.* https://doi.org/10.13140/RG.2.1.4929.1363.

Xie, B., Shen, Z., & Wang, K. (2021). Is preprint the future of science? A thirty year journey of online preprint services. *ArXiv:2102.09066 [Cs].* http://arxiv.org/abs/2102.09066.

ABOUT THE AUTHORS

Iratxe Puebla is Associate Director at ASAPbio, where she coordinates projects and community work in support of a productive use of preprints in the life sciences. Prior to ASAPbio, Iratxe worked in publishing for 16 years, where she developed a strong interest for open research and innovation in peer review.

Jessica Polka is Executive Director of ASAPbio, where she works to promote open and transparent research communication. Jessica performed postdoctoral research in the department of Systems Biology at Harvard Medical School following a PhD in Biochemistry from UCSF.

Oya Y. Rieger is a senior strategist on Ithaka S+R's Libraries, Scholarly Communication, and Museums team where she spearheads projects that reexamine the curation and preservation missions of cultural heritage organizations and explore sustainability models for open scholarship. Prior to Ithaka S+R, Oya served as Associate University Librarian at Cornell University Library overseeing digital scholarship and scholarly communication programs, including the oversight of arXiv.org.

CPSIA information can be obtained
at www.ICGtesting.com
Printed in the USA
LVHW060424230422
716950LV00020B/1183

9 781941 26947